AF441822

Polymer Characterization by Thermal Methods of Analysis

Edited by JEN CHIU

Plastics Department
E. I. du Pont de Nemours & Company
Wilmington, Delaware

Selected Papers Presented at the Symposium on Polymer Characterization by Thermal Methods of Analysis held during a joint meeting of the Polymer Chemistry and Rubber Chemistry Divisions of the American Chemical Society in Detroit, May 1-4, 1973. First published in *Journal of Macromolecular Science (Chemistry)*, Volume A8, Number 1, 1974.

MARCEL DEKKER, INC.   New York   1974

MARCEL DEKKER, INC.
305 East 45th Street, New York, New York 10017

LIBRARY OF CONGRESS CATALOG
CARD NUMBER: 74-77063
ISBN: 0-8247-6176-6

Current printing (last digit):
10 9 8 7 6 5 4 3 2 1

PRINTED IN THE UNITED STATES OF AMERICA

CONTENTS

CONTENTS

CONTRIBUTORS TO THIS VOLUME

(Number in parentheses indicate the pages on which the authors' contributions begin.)

EDWARD M. BARRALL II (135), IBM Research Laboratory, San José, California

J. M. BARTON (25), Materials Department, Royal Aircraft Establishment, Farnborough, Hants, England

D. M. CATES (165), Department of Textile Chemistry, North Carolina State University, Raleigh, North Carolina

PRONOY K. CHATTERJEE (191), Personal Products Company, Subsidiary of Johnson & Johnson, Milltown, New Jersey

JEN CHIU (3), Plastics Department, E. I. du Pont de Nemours & Company, Wilmington, Delaware

PETER CUKOR (105), GTE Laboratories, Inc. , Waltham, Massachusetts

BARBARA DAWSON (135), IBM Research Laboratory, San José, California

A. H. DiEDWARDO (119), Analytical Chemistry Department, Celanese Research Company, Summit, New Jersey

THOMAS J. GEDEMER (95), Research and Development, McGraw-Edison Power Systems Division, South Milwaukee, Wisconsin

B. M. GENDRON (175), General Electric Corporate, Research and Development, Schenectady, New York

J. K. GILLHAM (211), Polymer Materials Program, Department of Chemical Engineering, Princeton University, Princeton, New Jersey

IAN R. HARRISON (43), Department of Material Sciences, The Pennsylvania State University, University Park, Pennsylvania

W. T. HUMPHRIES (65), Department of Skin Biology, Johnson & Johnson Research, New Brunswick, New Jersey

PHILIP M. JAMES (135), IBM Research Laboratory, San José, California

B. L. JOESTEN (83), Chemicals and Plastics Division, Union Carbide Corporation, Bound Brook, New Jersey

N. W. JOHNSTON (83), Chemicals and Plastics Division, Union Carbide Corporation, Bound Brook, New Jersey

HIROTARO KAMBE (157), Institute of Space and Aeronautical Science, University of Tokyo, Komaba, Meguro-ku, Tokyo, Japan

TEIJI KATO (157), Institute of Space and Aeronautical Science, University of Tokyo, Komaba, Meguro-ku, Tokyo, Japan

E. KIRAN (211), Polymer Materials Program, Department of Chemical Engineering, Princeton University, Princeton, New Jersey

MASAKATSU KOCHI (157), Institute of Space and Aeronautical Science, University of Tokyo, Komaba, Meguro-ku, Tokyo, Japan

J. A. LOGAN (135), IBM Research Laboratory, San José, California

T. R. MANLEY (53), Department of Materials Science, Newcastle upon Tyne Polytechnic, Newcastle upon Tyne, England

JOHN J. MAURER (73), Enjay Polymer Laboratories, Linden, New Jersey

CARMINE PERSIANI (105), GTE Laboratories, Inc., Waltham, Massachusetts

A. R. SHULTZ (175), General Electric Corporate Research and Development, Schenectady, New York

W. K. WALSH (165), Department of Textile Chemistry, North Carolina State University, Raleigh, North Carolina

R. H. WILDNAUER (65), Department of Skin Biology, Johnson & Johnson Research, New Brunswick, New Jersey

PAUL E. WILLARD (33), FMC Corporation, Princeton, New Jersey

C. C. YAU (165)*, Department of Textile Chemistry, North Carolina State University, Raleigh, North Carolina

F. ZITOMER (119), Analytical Chemistry Department, Celanese Research Company, Summit, New Jersey

---

* Present address: Emory Industries, Inc., Cincinnati, Ohio

Introduction

This publication contains the papers presented at the Symposium on Polymer Characterization by Thermal Methods of Analysis held during a joint meeting of the Polymer Chemistry and Rubber Chemistry Divisions of the American Chemical Society in Detroit, May 1-4, 1973. Several papers have been revised and expanded from the oral versions. One paper, "Thermal Studies of Solubility and Diffusion of Antioxidants," by H. E. Bair, R. J. Roe, and C. Gieniewski, is not included in this volume because it constitutes part of another work to be published elsewhere. The contents of this paper, however, can be found in Polymer Preprints, 14(1), 530 (1973).

In recent years we have seen an enormous growth in the use of dynamic thermal methods in various branches of chemistry, particularly in polymer chemistry, and this trend is expected to continue for years to come. All these techniques are based on the measurement of changes in certain physical or chemical properties of a material continuously and automatically as a function of temperature. These properties include temperature, enthalpy, mass, volume, light transmission, electrical conductivity, mechanical and dielectric relaxations, sonic velocity, and chemical composition. The future expansion of the thermal analysis family seems unlimited.

The emphasis of this symposium is to describe recent advances in technology and new applications of such dynamic thermal techniques to polymer systems. The program starts with a state-of-the-art review of the whole field to provide background information and general references to orient the audience. The following research contributions are grouped approximately into three sessions according to techniques. The first group consists of papers mainly concerned with differential thermal analysis and differential scanning calorimetry. The papers presented at the second session are based on thermogravimetry or thermogravimetry coupled with gas chromatography and mass spectrometry. The third session deals with a number of thermal techniques such as thermo-mechanical analysis, thermo-optical analysis, thermoacoustical analysis, and thermal evolution analysis combined with pyrolysis-mass chromatography. Systems studied range from additives to plastics, fibers, elastomers, thermosetting resins, and aging skin. Applications of the techniques include study of phase transitions and morphology, evaluation and characterization of additives and polymers, and investigation of cross-linking and degradation reactions. Kinetics and mechanisms are discussed for several polymer systems.

I believe that the collection of papers in this volume is indicative of the current research efforts in dynamic thermal analysis, and should be a useful reference for those interested in polymer characterization by modern instrumental methods.

The editor wishes to express his gratitude to the authors for their contributions and for their active participation in the symposium. He also thanks Dr. J. J. Maurer for his help in presiding over the second session. It is a pleasure to acknowledge the support of the Instrument Products Division of the Du Pont Company to this symposium by sponsoring a banquet with the speakers.

Jen Chiu

Plastics Department
E. I du Pont de Nemours & Co.
Wilmington, Delaware 19898

Dynamic Thermal Analysis of Polymers.
An Overview

JEN CHIU

Plastics Department
E. I. du Pont de Nemours & Company
Wilmington, Delaware 19898

ABSTRACT

Dynamic thermal methods of analysis are becoming increasingly important for polymer characterization. These
techniques are based on measuring changes in certain
physical properties continuously and automatically upon
variation of the sample temperature. This presentation reviews thermal analysis in general, and briefly describes
the features and applications of several commonly used
thermal methods, such as differential thermal analysis,
differential scanning calorimetry, thermogravimetry, thermomechanical analysis, electrothermal analysis, thermo-
optical analysis, and thermal evolution analysis. General
references are given for the various techniques.

INTRODUCTION

Thermal methods used for polymer characterization can be traced
to ancient times beyond memory, if we include all those techniques
associated with a temperature measurement or a thermal treatment;
for instance, a burning test to distinguish fibers, a capillary melting

point determination, or a dripping test. It will also be beyond the
scope of this presentation to include such thermal methods as
pyrolysis-gas chromatography, -mass spectrometry, and -infrared
spectroscopy; hot-stage type of microscopy; classical calorimetry;
or large-scale physical testing with temperature variation. We are
here concerned with "dynamic thermal analysis," a family of rapidly
growing analytical techniques based on monitoring changes in a
physical property continuously and automatically as a function of
temperature or as a function of time at a controlled temperature.

Most of the thermal techniques are old in principle but new in
instrumentation. Besides the use of small sample size to achieve
high resolution of thermal events, a recent trend in instrumentation
is to provide modules to perform various functions, and to share
some common components such as temperature programmer,
recorder, atmosphere control, and even amplifiers. The present
fluorishing use of dynamic thermal methods in polymer chemistry
started around 1961 when instruments with adequate resolution for
polymer studies became available. Presently, at least 30 manufacturers are marketing thermal analyzers [1, 2]. The number of
publications related to thermal analysis in basic chemical journals
has been increasing at an exponential rate, and is now estimated at
more than 5000 annually, 20% of which are probably associated with
polymers [1, 3]. Two journals, Thermochimica Acta [4] and Journal
of Thermal Analysis [5], are devoted to thermal analysis exclusively.
Thermal analysis abstracts [6] are also becoming available.

The International Confederation for Thermal Analysis (ICTA) has
held three conferences since its formation in 1965, the proceedings of
which have been published [7-9]. Professional societies for thermal
analysis have been formed in Japan, England, Italy, the Soviet Union,
and other nations. Most thermal analysts in the United States join the
North American Thermal Analysis Society (NATAS). The proceedings
of the third annual NATAS meeting held in February 1972 have been
published [10]. Several symposiums on thermal analysis were
sponsored by various divisions of the American Chemical Society during national meetings in 1963 [11], 1965 [12], 1968 [13], and 1970
[14, 15]. These conferences contributed greatly to the exchange of
research results among various workers. Interlaboratory tests are
being conducted on several thermal techniques by ICTA, ASTM,
International Organization for Standardization (ISO), and other national
groups, and some results have been reported [16-21]. In collabora-
tion with ICTA, the U.S. National Bureau of Standards has issued three
sets of temperature standards (NBS-ICTA Standard Reference Materials
758, 759, and 760) for DTA in the three temperature ranges 125-435,
295-675, and 570-950°C, respectively. The program leading to the
issuance of such standards and current activities of various organiza-
tions on standardization have been described [17, 22]. Recommend-
tions on certain aspects of nomenclature in thermal analysis have

been published by ICTA [23], and should be consulted in preparing publications.

General descriptions of various thermal methods are provided in books by Wendlant [24] and Garn [25]. Fundamental developments in thermal analysis are reviewed biennially by Murphy [26].

THERMAL ANALYSIS OF POLYMERS

Recent developments in the use of thermal methods for polymer characterization are reviewed biennially by Mitchell and Chiu [27]. Two volumes in a series of books on techniques and methods of polymer evaluation edited by Slade and Jenkins [28] are devoted to thermal techniques. Other review articles on thermal analysis of polymers in general include those by Ehlers and Fisch [29], Kambe [30], Miller [31], Manche and Carroll [32], and Nakamura [33]. Comprehensive thermal analysis data, mainly DTA and TGA, are included in books by Frazer [34] and Korshak [35].

The applications of thermal analysis to polymer chemistry can be briefly summarized.

Monomers	polymerization curing	
Solvents	depolymerization	Polymers
Additives	decomposition	

Polymerization and curing of monomeric systems can be studied by many thermal techniques with respect to enthalpy changes or kinetics and mechanism. In the opposite direction, depolymerization and decomposition of polymers under controlled conditions can be investigated and reaction products analyzed. For polymers, thermal techniques provide information on melting, crystallization, softening, shrinking, glass transition and other relaxations, crystalline transitions, adsorption, dissolution, thermal conductivity, crystallinity, heat capacity, dielectric properties, electrical conductivity, modulus, thermal expansion, thermal stability, oxidation and aging, chemical composition and structure, and chemical reactions. Thermal techniques are also used for studies of the components making polymers including monomers, solvents, additives, and initiators. Properties frequently studied include melting, freezing, crystalline transitions, glass transition, magnetic transition, boiling, sublimation, vapor pressure, heat capacity, purity, adsorption, electric behavior, thermal stability, solvation, complexation, oxidation and combustion, and other chemical reactions.

It is impossible and unnecessary to elaborate on these applications in this brief review. Many unique capabilities of thermal analysis for polymer characterization will be presented in this symposium.

However, some general comments will be made here on each of the more commonly used thermal techniques; namely, differential thermal analysis (DTA), differential scanning calorimetry (DSC), thermogravimetry (TG), thermomechanical analysis (TMA), electrothermal analysis (ETA), thermo-optical analysis (OTA), and thermal evolution analysis (TEA).

DIFFERENTIAL THERMAL ANALYSIS (DTA)/ DIFFERENTIAL SCANNING CALORIMETRY (DSC)

DTA and DSC are techniques monitoring enthalpy changes of a sample as a function of temperature. DSC or dynamic differential calorimetry (DDC) is considered a quantitative version of DTA, and therefore not distinguished in most discussions. DTA is probably the first dynamic thermal technique to arouse the enthusiasm of polymer chemists. General reviews on DTA include those by Smothers and Chiang [36], Mackenzie [37], Murphy [38], Wunderlich [39], and Schultze [40]. Reviews devoted to DTA of polymers have been written by Ke [41], Teitel'baum and Anoshina [42], Smith [43], Kanetsuna [44], and Barrall and Johnson [45].

DTA/DSC is mainly used for studies of physical transitions, although reactions such as degradation [46], polymerization [47, 48], oxidation [49-51], vulcanization [52], and chemical reactions [53] can readily be studied. Important applications of DTA/DSC to polymers and intermediates also include heat capacity [54, 55], crystallinity [56], and purity [57-62] measurements. Melting temperatures are frequently used for polymer identification on the basis that each polymer shows its characteristic melting endotherm in an incompatible mixture. This was demonstrated [63] by DTA of a physical mixture of seven commercial polymers; namely, high-pressure polyethylene, low-pressure polyethylene, polypropylene, polyoxymethylene, nylon 6, nylon 66, and polytetrafluoroethylene (Fig. 1). The melting peak area is a measure of the heat of fusion of the polymer, and can be used to determine the amount of the polymer present.

Qualitative analysis of polymers relies heavily on the resolution of thermal events. High resolution becomes possible only when a small sample is used. This is shown in Fig. 2 by DTA of a blend of linear polyethylene and isotactic polypropylene in a ratio of 25:75 by weight [64]. The two melting endotherms are completely resolved when a sample size of 2 mg is used. Another factor is heating rate. In general, a slower heating rate provides better resolution. Resolution of DTA peaks, R, can be defined, as shown in Fig. 3, in a manner similar to that in chromatography as the distance, $T_2 - T_1$, between the two peak maxima divided by the sum of the two half-peak widths, $t_2 + t_1$. The effect of sample size and scanning rate on resolution can then be

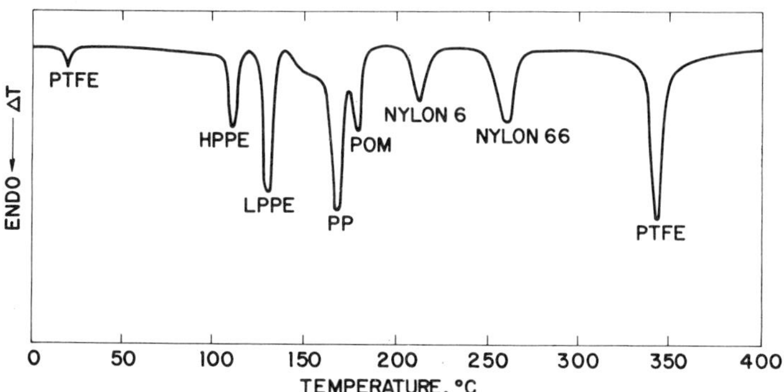

FIG. 1. DTA of seven component polymer mixture. Sample weight, 8 mg; heating rate, 15°C/min.

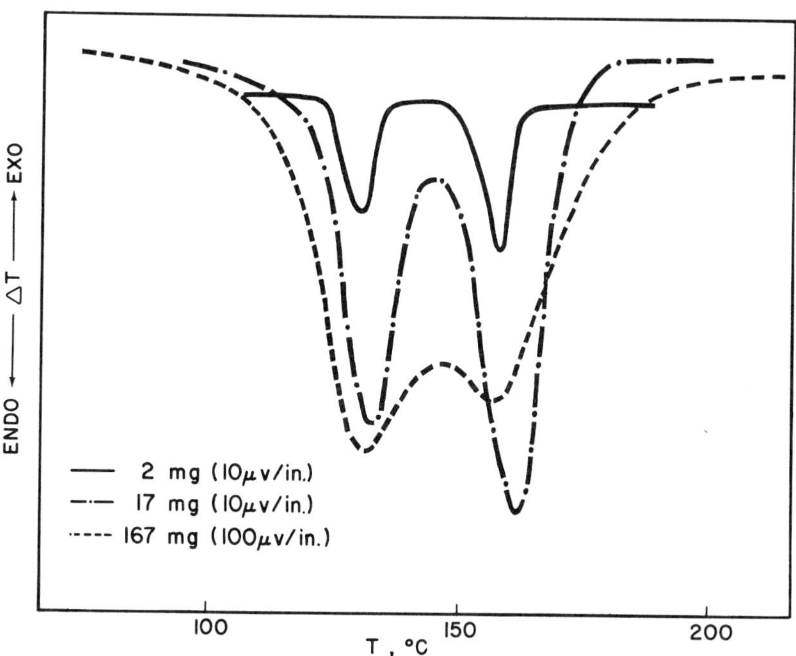

FIG. 2. DTA of 25:75 polyethylene:polypropylene blend. Heating rate, 45°C/min.

represented by the relationship

$$RS^a W^b = C$$

where S and W represent scanning rate and weight of sample, respectively, and a, b, and C are empirical constants. In the meantime, the effect of sample size and scanning rate on the transition temperature can be represented by

$$T = T_0 \pm k_1 S \pm k_2 W$$

where T is the observed transition temperature, T_0 is the equilibrium transition temperature, and k_1 and k_2 are instrumental constants. Plus signs are used for heating and minus signs for cooling. The above two relationships have been tested with experiments on resolution of the two closely spaced transitions of n-dotriacontane, chain rotation and fusion. The results obtained using a Perkin-Elmer DSC-1B instrument are shown in Tables 1 and 2 and plotted in Figs. 4-7. Infinite resolution is thus obtained at zero scanning rate or zero sample size. True transition temperatures are obtained when both scanning rate and sample size approach zero. We obtained the thermogram for dotriacontane shown in Fig. 8(C) by using a sample size of 0.1 mg, a heating rate of 10°C/min, and a chart speed of 6 in./min. A resolution of 4.7 was achieved with peak temperatures at 65.1 and 70.4°C for rotation and fusion, respectively. This resolution contrasts dramatically with the scan in Fig. 8(B) obtained under normal operating conditions and the scan in Fig. 8(A) obtained under somewhat extreme conditions.

A question often raised by polymer chemists is how to determine the melting point of a sample from its melting endotherm obtained by DTA or DSC. For metals and monomeric substances, the extrapolated onset temperature, found by extrapolating the base line, prior to the peak, and the leading side of the peak to their intersection is the closest

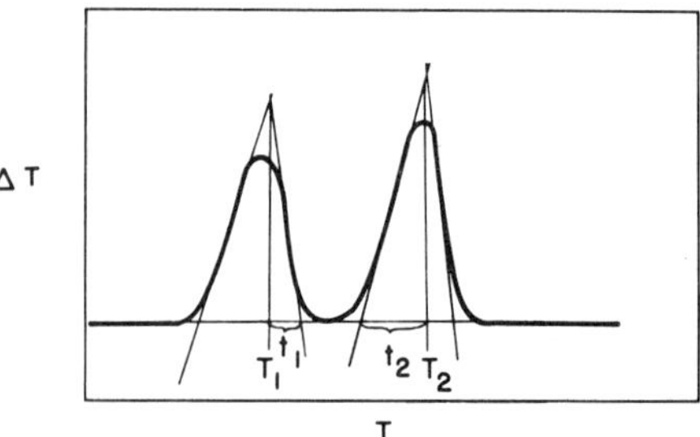

FIG. 3. Resolution.

TABLE 1. Sample Size Effect (scanning rate = 40°C/min)

Sample size (mg)	Heating			Cooling		
	$T_1(°C)$	$T_2(°C)$	[R]	$T_1(°C)$	$T_2(°C)$	[R]
10.0	73.0	82.0	0.53	-	-	-
5.0	70.8	78.0	0.71	54.8	60.0	0.43
1.0	68.7	74.8	1.10	57.0	63.2	1.41
0.2	66.8	73.0	1.44	57.0	63.0	1.90
0.05	67.5	72.9	2.18	57.3	63.3	2.87

TABLE 2. Scanning Rate Effect (sample size = 5.0 mg)

S (°C/min)	Heating			Cooling		
	$T_1(°C)$	$T_2(°C)$	[R]	$T_1(°C)$	$T_2(°C)$	[R]
40	71.0	77.0	0.78	55.0	60.0	0.41
20	68.0	74.2	1.04	59.1	64.3	0.96
10	66.5	72.8	1.59	60.9	65.8	1.70
5	65.8	71.3	2.00	62.0	66.8	2.86
2.5	65.2	70.6	2.84	62.4	67.6	3.84
1.25	65.4	70.8	3.76	63.2	68.4	6.58

to the equilibrium transition temperature [17]. However, this is in-applicable for polymers. Melting of a polymer occurs over a broad temperature range, and the melting point observed varies greatly with the sensitivity of the method used for the determination. The use of the deviation point or the extrapolated onset point in a DTA/DSC scan could cause great chaos. An interlaboratory study of melting point methods for polyamides, sponsored by ISO, was made on seven nylons using Fisher-Johns method of ASTM D 789, Kofler hot stage method of ASTM D117, Leitz polarizing microscope, Antes capillary, Culatti capillary, x-ray diffraction, and DTA. The results reported by laboratories in Italy, France, Germany, The Netherlands, and the United States are summarized and shown in Fig. 9 [21]. The spread of the data even within similar methods is indeed disappointing. However, it is of interest to note that the DTA extrapolated end and end tempera-tures are close to those by x-ray diffraction, and the DTA peak temper-atures are essentially the averages of the other methods. Since the

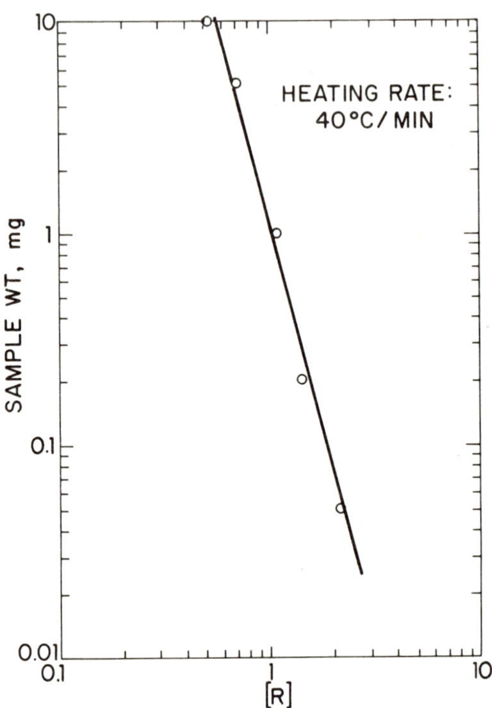

FIG. 4. Sample size effect on resolution on n-dotriacontane transitions.

x-ray method measures a more definitive temperature of disappearance of the last traces of crystallinity, and is more objective and sensitive than the optical methods, the extrapolated end point or the end point seems to be the temperature of choice as the DTA melting point of a polymer sample.

For amorphous polymers, glass transitions can sometimes be used for qualitative and quantitative analyses [65]. More recent developments in DTA/DSC include high-pressure operations [66-68], simultaneous DTA/DSC and electrical conductivity measurements [69-71] simultaneous DTA and photothermal analysis [72], thermal diffusivity determination [73], combined DTA/DSC-mass spectrometry [74, 75], and computer reduction of DSC data [76-78]. Computer interface and programs are also commercially available.

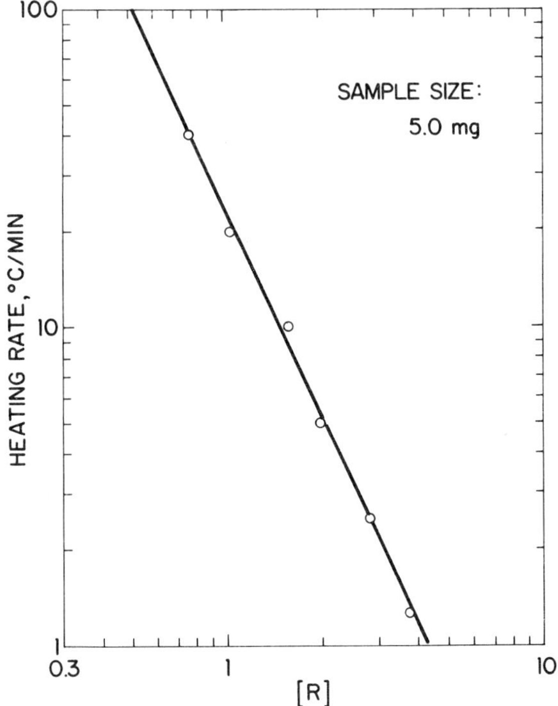

FIG. 5. Heating rate effect on resolution of n-dotriacontane transitions.

THERMOGRAVIMETRY (TG)

Thermogravimetry or thermogravimetric analysis (TGA) monitors the weight changes in a sample as a function of temperature. General reviews on TG of polymers include those by Chiu [79], Anderson [80], Reich and Levi [81], Zimmermann and Budisch [82], Reich [83], and Reich and Stivala [84].

TG is considered the most important method for studying polymer stability. However, there are other unique applications of this technique not well recognized, such as determination of additives [79, 85] and characterization of polymer blends and copolymers [79, 86, 87]. TG has also been used for studies on flame retardance [88-93], cross-linking [94, 95], track resistance [96], and thermal life [97, 98].

Derivative thermogravimetry (DTG) shows improved resolution of thermal events and precise comparison of thermal stability [99-101].

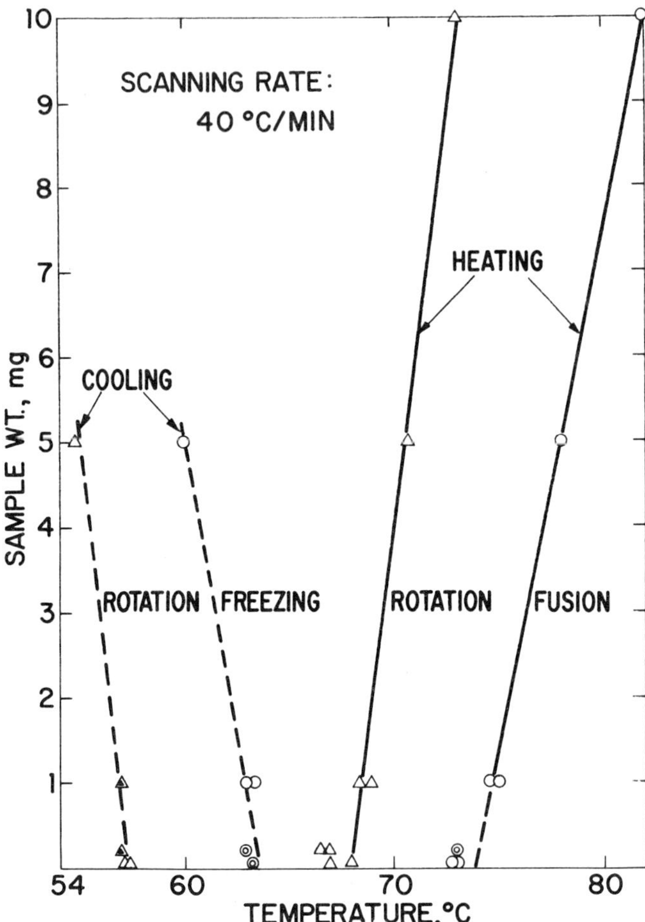

FIG. 6. Sample size effect on peak temperatures of n-dotriacontane transitions.

Kinetic studies by TG have been reviewed [102, 103]. Isothermal and dynamic TG methods for kinetic studies of polymer degradation have been compared [104, 105]. The dynamic methods provide more rapid measurements over a wide temperature range, whereas the isothermal technique is more reliable and applicable to most systems. Computers have been used for processing TG data, particularly in kinetic studies [106-114].

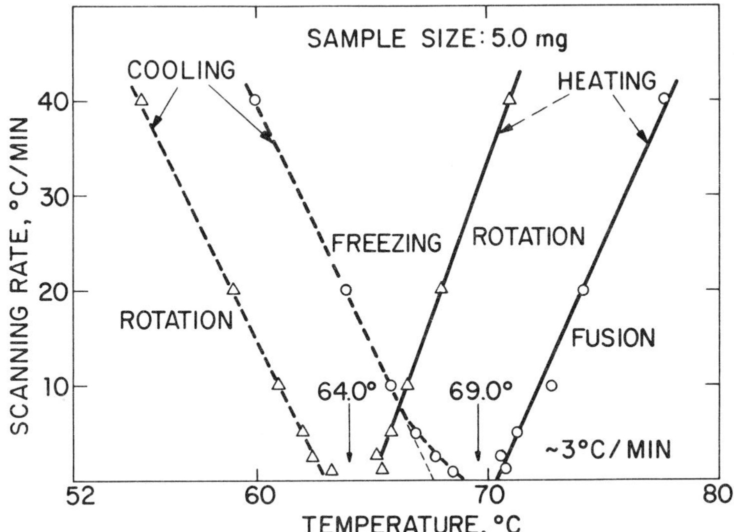

FIG. 7. Scanning rate effect on peak temperatures of n-dotriacontane transitions.

TG is basically a means for studying behavior, and not an absolute identification tool. For investigation of complex systems, it is often desirable to couple TG with other analytical techniques such as DTA, DTA-ETA [115], DTA-TMA [116], TOA [117], gas chromatography (GC) [118], mass spectrometry (MS) [119-121], GC-MS [122, 123], and GC-IR [124].

THERMOMECHANICAL ANALYSIS (TMA)

Thermomechanical analysis is a name broadly used to describe a group of thermal techniques which measure the mechanical strength of a material as a function of temperature. Such techniques are useful in obtaining information on softening, modulus, shrinkage, glass transition, segmental motions, and expansion coefficients. Commercial instruments are available to allow dynamic measurements performed in expansion, compression, and extension modes [125-127]. The TMA method for relating modulus and temperature has been shown to be comparable to ASTM tests such as deflection temperature under load, VICAT softening temperature, and Clash-Berg T_F [128].

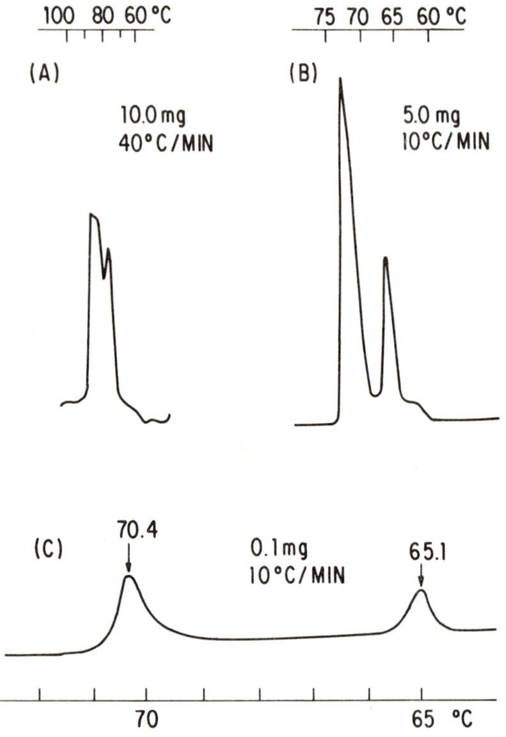

FIG. 8. DSC scans of n-dotriacontane.

TMA has been used for studies of phase transitions [129-132], swelling and dissolution [133, 134], curing [135, 136], and rheological compatibility of polymer blends [137].

Another thermomechanical technique, named torsional braid analysis (TBA) by Gillham [138], is an extension of the conventional torsion pendulum method for dynamic mechanical studies. The method employs a glass braid substrate, impregnated with a solution of the material under investigation, and subjected to free torsional oscillations under programmed temperature conditions. The frequency and decay of the oscillating pendulum provide information on the modulus and mechanical damping of the sample, allowing studies of polymer transitions, resin curing, and degradations. TBA has been recently used for determination of the effectiveness of polymer additives [139]. A similar method using a dynamic spring has been used for following cross-linking reactions with higher sensitivity than TBA [140].

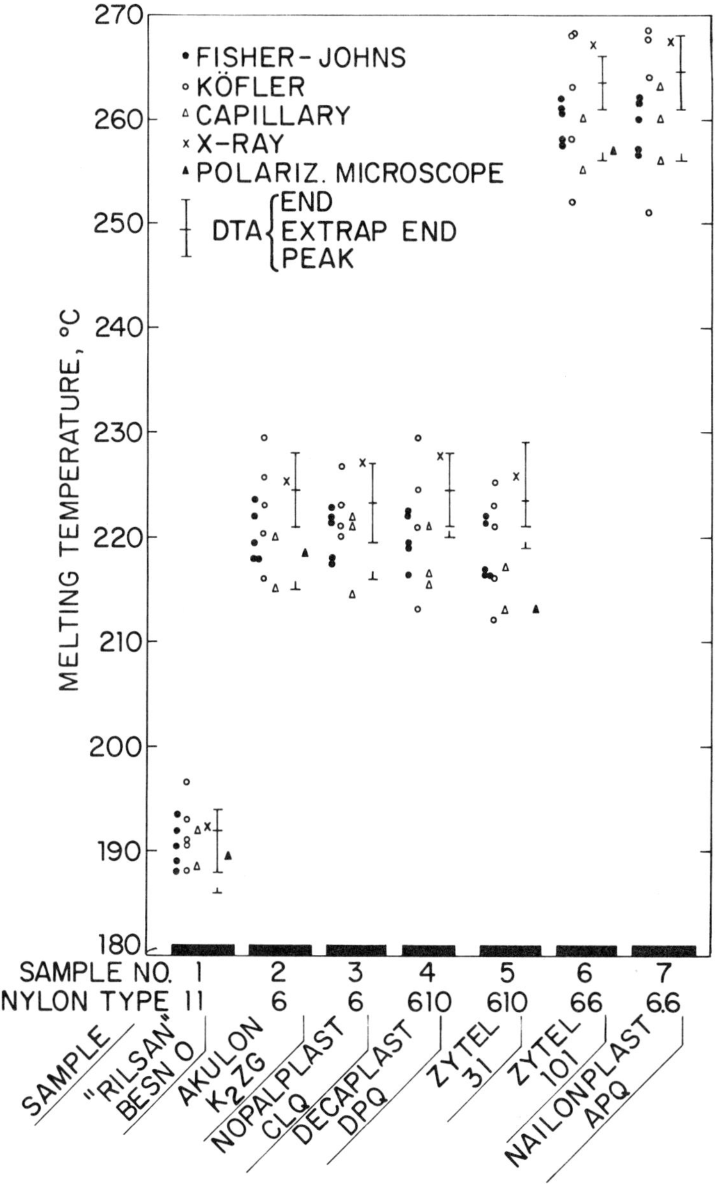

FIG. 9. Melting points of polyamides by different methods.

ELECTROTHERMAL ANALYSIS (ETA)

Electrothermal analysis is defined as a thermoanalytical technique for studying the nature and behavior of a material by monitoring the changes in resistivity or dielectric parameters automatically and continuously as a function of temperature. Thus electrical properties can be used in polymer characterization in a manner similar to DTA or TMA. Although this technique has not yet been widely used, potential applications of ETA to polymer problems evidently include studies of physical transitions and segmental relaxations, analysis of additives and contaminants, elucidation of structures, identification of polymer blends and copolymers, and investigation of polymerization, curing, and degradation reactions.

Principles and applications of direct current (dc) ETA or resistivity measurements have been reviewed [141, 142]. Recent developments include simultaneous DSC-ETA [71], DTA-ETA [69, 70, 143], and DTA-TG-ETA [115].

Dielectric measurements performed as a function of temperature have shown promise in elucidation of structure and stereoregularity of acrylics [144]. Sacher [145] used a continuously nulling capacitance bridge to obtain both capacitance and dissipation factor vs temperature curves for a variety of polymers. Hedvig, Kisbenyi, et al. used the dielectric method to study degradation [146], efficiency and compatibility of plasticizers [147], and curing [148]. Wrasidlo and co-workers [149] and Chiu [150] developed apparatus for dynamic dielectric measurements, and applied the technique to several polymer systems. The correlation among ETA, DTA, and other methods was discussed.

THERMO-OPTICAL ANALYSIS (TOA)

Thermal depolarization analysis (TDA), depolarized light intensity analysis (DLI), and thermo-optical analysis (TOA) are names used to describe thermal techniques based on measuring light intensity as a function of temperature. Generally, this technique is more sensitive than DTA in detecting partial melting, pre-melting, and recrystallization phenomena. The use of changes in optical properties for studying fusion has been discussed [151]. A general discussion of the use of thermal optical measurements for polymer characterization has been presented [152-154]. Major applications of TOA include studies on glass transition [155], melting [156], mobility transitions in polymer blends [157], and smoke generation [117]. A technique based on light emanation from sample upon heating has also been reported [72].

THERMAL EVOLUTION ANALYSIS (TEA)

Evolved gas analysis (EGA) and evolved gas detection (EGD) generally refer to techniques which determine the nature or amount of volatiles formed, or detect whether or not a volatile product is formed. Here the name thermal evolution analysis is coined to include techniques which monitor continuously the amount of volatiles thermally evolved from the sample upon programmed heating. The resulting record is essentially equivalent to a weight loss curve. In fact, this technique provides information similar to that obtained by TG. The advantages of TEA over TG are potentially higher sensitivity, less expensive equipment, and more flexibility in sample handling. Commercial units are presently available.

Transducers used in TEA include vacuum gauge [158], flame ionization detector [159, 160], and thermal conductivity detector [161]. These techniques have shown promise in studies of thermal degradation, analysis of trace impurities and additives, and elucidation of polymer structures. Another potential application is in the area of vapor pressure measurements for flame retardants and other additives in polymer systems [162].

CONCLUSION

The above discussion provides a brief overview of the various dynamic thermal methods more commonly used today. More detailed information can be found in the general references given. Thermal analysis is now generally recognized as one of the basic analytical tools for polymer characterization. Its future expansion can only be limited by the usefulness of the property of the material to be measured. With the rapid development and improvement of commercial instrumentation, and the increasing awareness of the capabilities of thermal techniques by industries and research institutes, the continued growth of thermal analysis is expected for years to come.

REFERENCES

[1] Chem. Eng. News, 47, 46 (August 18, 1969).
[2] "Labguide," Anal. Chem., 44(10) (1972).
[3] D. A. Teetsel and D. W. Levi, PLASTEC (Plastics Technology Evaluation Center) Note, 7 (1963); 10 (1966); 20 (1969).
[4] W. W. Wendlandt, ed., Thermochimica Acta, Elsevier, Amsterdam 1st issue, 1970.

[5]   E. Buzagh and J. Simon, eds., Journal of Thermal Analysis, Heyden & Son, London, 1st issue, 1969.

[6]   J. P. Redfern, ed., Thermal Analysis Abstracts, Heyden & Son, London, 1st issue, 1972.

[7]   J. P. Redfern, ed., Thermal Analysis, Macmillan, London, 1965.

[8]   R. F. Schwenker, Jr. and P. D. Garn, eds., Thermal Analysis, Academic, New York, 1969, 2 vols.

[9]   H. G. Wiedemann, ed., Thermal Analysis, Birkhäuser, Basel, Switzerland, 1972, 3 vols.

[10] Thermochimica Acta, 4(3-5), (1972).

[11] B. Ke, ed., Thermal Analysis of High Polymers, J. Polym. Sci., C(6), (1964).

[12] R. F. Schwenker, Jr., ed., Thermoanalysis of Fibers and Fiber-forming Polymers, Appl. Polym. Symp., 2, (1966).

[13] R. S. Porter and J. F. Johnson, eds., Analytical Calorimetry, Plenum, New York, 1968.

[14] R. S. Porter and J. F. Johnson, eds., Analytical Calorimetry, Vol. 2, Plenum, New York, 1970.

[15] Thermochimica Acta, 1(2-6), (1970); 2(1), (1971).

[16] H. G. McAdie, in Ref. 8, p. 1499.

[17] O. Menis and J. T. Sterling, Nat. Bur. Stand. Spec. Publ., 338, 61 (1970).

[18] R. D. Berg, Soc. Plast. Eng., Tech. Pap., 17, 163 (1971).

[19] T. J. Gedemer, Soc. Plast. Eng., Tech. Pap., 18 (Pt. 1), 361 (1972).

[20] Z. Osawa, Bunseki Kagaku, 19 (3), 415 (1970).

[21] ISO interlaboratory study of melting point methods for preparation of ISO recommendation 1218, 1970. Data available from Secretariat, ISO/TC 61, American National Standards Institute, New York.

[22] H. G. McAdie, P. D. Garn, and O. Menis, Nat. Bur. Stand. Spec. Publ., 260-40, (1972).

[23] R. C. Mackenzie, Talanta, 16, 1227 (1969); 19, 1079 (1972).

[24] W. W. Wendlandt, Thermal Methods of Analysis, Wiley (Interscience), New York, 1964.

[25] P. D. Garn, Thermoanalytical Methods of Investigation, Academic, New York, 1965.

[26] C. B. Murphy, Anal. Chem., 36, 347R (1964); 38, 443R (1966); 40, 381R (1968); 42, 268R (1970); 44, 513R (1972).

[27] J. Mitchell, Jr. and J. Chiu, Anal. Chem., 41, 248R (1969); 43, 267R (1971); 45, 273R (1973).

[28] P. E. Slade, Jr. and L. T. Jenkins, eds., Techniques and Methods of Polymer Evaluation, Dekker, New York, Vol. 1, 1966; Vol. 2 1970.

[29] G. F. L. Ehlers and K. R. Fisch, Appl. Polym. Symp., 8, 171 (1969).

[30] H. Kambe, Kobunshi, 17(200), 1193 (1968).

[31] G. W. Miller, Appl. Polym. Symp., 10, 35 (1969).
[32] E. P. Manche and B. Carroll, Phys. Methods Macromol. Chem.,
      2, 239 (1972).
[33] S. Nakamura, Nippon Gomu Kyokaishi, 42(8), 592 (1969).
[34] A. H. Frazer, High Temperature Resistant Polymers, Wiley
      (Interscience), New York, 1968.
[35] V. V. Korshak, Heat-Resistant Polymers, also The Chemical
      Structure and Thermal Characteristics of Polymers, Israel
      Program for Scientific Translations, Jerusalem, 1971.
[36] W. J. Smothers and Y. Chiang, Handbook of Differential Thermal
      Analysis, Chemical Publishing Co., New York, 1966.
[37] R. C. Mackenzie, ed., Differential Thermal Analysis, Academic,
      New York, Vol. 1, 1970; Vol. 2, 1972.
[38] C. B. Murphy, Treatise on Analytical Chemistry, Part 1, Wiley
      (Interscience), New York, 1968, p. 5243.
[39] B. Wunderlich, in Physical Methods of Chemistry, Pt. 5, (A.
      Weissberger and B. W. Rossiter, eds.), Wiley, New York, 1971.
[40] D. Schultze, Differential-thermoanalyse, 2nd ed., Verlag Chemie,
      Weinheim, 1971.
[41] B. Ke, in Newer Methods of Polymer Characterization (B. Ke,
      ed.), Wiley (Interscience), New York, 1964.
[42] B. Ya. Teitel'baum and N. P. Anoshina, Russ. Chem. Rev., 36, 37
      (1967).
[43] D. A. Smith, Rubber J., 150(4), 21(1968).
[44] H. Kanetsuna, Kobunshi, 17(200), 1087 (1968).
[45] E. M. Barrall, II and J. F. Johnson, Crit. Rev. Anal. Chem., 2(1),
      105 (1971).
[46] L. Reich, Macromol. Rev., 3, 49 (1968).
[47] J. Chiu, in Ref. 14, p. 171.
[48] K. E. J. Barrett and H. R. Thomas, Brit. Polym. J., 2(1-2), 45
      (1970).
[49] A. Rudin, H. P. Schreiber, and M. H. Waldman, Ind. Eng. Chem.,
      53, 137 (1961).
[50] E. Wiesener, Faserforsch. Textiltech., 21(12), 514 (1970).
[51] B. Wargotz, Amer. Chem. Soc., Org. Coatings Plastics Chem.
      Div. Preprints, 31(1), 639 (1971).
[52] M. L. Bhaumik, A. K. Sircar, and D. Banerjee, J. Appl. Polym.
      Sci., 4, 366 (1960).
[53] J. Chiu, Anal. Chem., 34, 1841 (1962).
[54] B. Wunderlich, J. Phys. Chem., 69, 2078 (1965).
[55] M. J. O'Neill, Anal. Chem., 38, 1331 (1966).
[56] A. P. Gray, Thermochim. Acta, 1, 563 (1970).
[57] C. Plato, Anal. Chem., 44, 1531 (1972).
[58] D. L. Sondack, Ibid., 44, 888 (1972).
[59] E. M. Barrall, II and R. D. Diller, Thermochim. Acta, 1(6),
      509 (1970).
[60] E. F. Joy, J. D. Bonn, and A. J. Bernard, Jr., Ibid., 2(1), 57
      (1971).

[61] R. Schmacher and B. Felder, Fresenius' Z. Anal. Chem., 254(4), 265 (1971).

[62] H. Staub and W. Perron, Pittsburgh Conference on Analytical Chemistry and Applied Spectroscopy, Cleveland, Ohio, March 6, 1972.

[63] J. Chiu, Du Pont Thermogram, 2(3), 9 (1965).

[64] J. Chiu, Ibid., 1 (1), 1 (1964).

[65] H. E. Bair, in Ref. 14, p. 51.

[66] T. Davidson and B. Wunderlich, J. Polym. Sci., A-2, 7, 377 (1969).

[67] P. F. Levy, G. Nieuweboer, and L. C. Semanski, Thermochim. Acta, 1, 429 (1970).

[68] M. Yasuniwa, C. Nakafuku, and T. Takemura, Kyushu Daigaku Kogaku Shuho, 45(1), 1 (1972).

[69] J. Chiu, J. Polym. Sci., C, 8, 27 (1965).

[70] D. J. David, Thermochim. Acta, 1, 277 (1970).

[71] R. W. Carroll and R. V. Mangravite, in Ref. 8, p. 189.

[72] D. J. David, Thermochim. Acta, 3, 277 (1972).

[73] S. Guillerur, Plastics Mod. Elastomers, 20(9), 116 (1968).

[74] G. Dugan, J. D. McCarty, and R. J. Friant, in Ref. 14, p. 417.

[75] R. J. Gaymans, K. A. Hodd, and W. A. Holmes-Walker, Polymer, 12(10), 602 (1971).

[76] E. A. Dorko, R. W. Crossley, and R. L. Diggs, in Ref. 14, p. 429.

[77] R. B. Prime, in Ref. 14, p. 201.

[78] L. J. Taylor and S. W. Watson, Amer. Chem. Soc., Div. Polym. Chem. Preprints, 12(1), 457 (1971).

[79] J. Chiu, Appl. Polym. Symp., 2, 25 (1966).

[80] H. C. Anderson, Tech. Methods Polym. Eval., 1, 87 (1966).

[81] L. Reich and D. W. Levi, Macromol. Rev., 1, 173 (1967).

[82] H. Zimmermann and J. Budisch, J. Thermal Anal., 1 (1), 107 (1969).

[83] L. Reich, Encyclopedia of Polymer Science and Technology, Vol. 14, Wiley (Interscience), New York, 1971, p. 1.

[84] L. Reich and S. S. Stivala, Elements of Polymer Degradation, McGraw-Hill, New York, 1971.

[85] J. J. Maurer, Rubber Age (New York), 102(2), 47 (1970); also in Ref. 8, p. 373.

[86] R. Boni, B. Filippi, L. Ciceri, and E. Peggion, Biopolymers, 9(12), 1539 (1970).

[87] G. J. Mol, Plastic Des. Process., 11(12), 20 (1971).

[88] R. M. Perkins, G. L. Drake, Jr., and W. A. Reeves, J. Appl. Polym. Sci., 10, 1041 (1966).

[89] J. K. Backus, D. L. Bernard, W. C. Darr, and J. H. Saunders, Ibid., 12, 1053 (1968).

[90] G. S. Learmonth and D. G. Thwaite, Brit. Polym. J., 2(3), 104 (1970).

[91] D. B. Parrish and R. M. Pruitt, J. Cell. Plastics, 5(6), 348 (1969).

[92] J. K. Smith, H. R. Rawls, M. S. Felder, and E. Klein, Text. Res. J., 40(3), 211 (1970).

[93] P. E. Ingham, J. Appl. Polym. Sci., 15, 3025 (1971).

[94] G. F. D'Alelio, D. M. Feigl, H. E. Kieffer, and R. K. Mehta, J. Macromol. Sci., A2(6), 1223 (1968).

[95] R. Y. Wen, L. F. Sonnabend, and R. Eddy, J. Macromol. Sci., A3(3), 471 (1969).

[96] G. H. Jaegers and T. J. Gedemer, Mod. Plast., 48(4), 110 (1971).

[97] D. J. Toop, IEEE Trans. Elec. Insul., 6(1), 2 (1971).

[98] H. A. Papazian, Thermochim. Acta, 4, 81 (1972); J. Appl. Polym. Sci., 16, 2503 (1972).

[99] O. Vogl, V. Ivansons, H. C. Miller, and H. W. Williams, J. Macromol. Sci., A2, 175 (1968).

[100] S. E. Gordon and R. C. Caballero, in Ref. 14, p. 155.

[101] T. J. Gedemer, Soc. Plastic Eng., Tech. Pap., 18 (Pt. 1), 105 (1972).

[102] C. D. Doyle, Tech. Methods Polym. Eval., 1, 113 (1966).

[103] J. H. Flynn and L. A. Wall, J. Res. Nat. Bur. Stand., A, 70, 487 (1966).

[104] J. R. MacCallum and J. Tanner, Eur. Polym. J., 6, 907 (1970); Nature, 225, 1127 (1970).

[105] R. Audebert, C. Aubineau, and J. Chiu, Phys. Physicochim. Biol., 66(3), 414 (1969); 67(3), 617 (1970).

[106] M. Beranek, Silikaty, 10, 93 (1966).

[107] J. M. Schempf, F. E. Freeberg, D. J. Royer, and F. M. Angeloni, Anal. Chem., 38, 520 (1966).

[108] R. W. Farmer, U.S. Air Force Materials Laboratory Tech. Rept., AFML-TR-65-246, Part III (1970); Thermochim. Acta, 4, 203, 223 (1972).

[109] I. J. Goldfarb, Amer. Chem. Soc., Div. Polym. Chem. Preprints, 10(2), 1289 (1969).

[110] I. J. Goldfarb, R. McGuchan, and A. C. Meeks, U.S. Air Force Materials Laboratory Tech. Rept., AFML-TR-68-181, Parts I and II (1968).

[111] S. R. Hobart and C. H. Mack, in Ref. 8, p. 571.

[112] N. J. Olson, U.S. Air Force Materials Laboratory Tech. Rept., AFML-TR-70-262 (1970).

[113] S. R. Urzendowski and A. H. Guenther, J. Therm. Anal., 3(4), 379 (1971).

[114] N. W. Johnston and B. L. Joesten, J. Polym. Sci., A-1, 10, 1271 (1972).

[115] J. Chiu, Anal. Chem., 39, 861 (1967).

[116] F. Paulik, J. Paulik, and L. Erdey, Talanta, 13, 1405 (1966).

[117] A. A. Loehr and P. F. Levy, Amer. Lab., 4(1), 11 (1972).

[118] J. Chiu, Anal. Chem., 40, 1516 (1968); Thermochim. Acta, 1(3), 231 (1970).

[119] F. Zitomer, Anal. Chem., 40, 1091 (1968).

[120] D. E. Wilson and F. M. Hamaker, in Ref. 8, p. 517.

[121] D. L. Geiger and G. A. Kleineberg, 20th Annual Conference on Mass Spectroscopy and Allied Topics, Dallas, Texas, June 4, 1972.

[122] T. L. Chang and T. E. Mead, Anal. Chem., 43, 534 (1971).

[123] A. H. DiEdwards and F. Zitomer, 17th Annual Conference on Mass Spectroscopy and Allied Topics, Dallas, Texas, May 18, 1969.

[124] P. Cukor and E. W. Lanning, J. Chromatogr. Sci., 9, 487 (1971).

[125] P. F. Levy and R. W. Tabeling, U.S. Patent 3,474,658 (October 28, 1969).

[126] S. H. Byrne, Jr. and D. L. Casey, U.S. Patent 3,583,208 (June 8, 1971).

[127] H. I. Hill, U.S. Patent 3,589,167 (June 29, 1971).

[128] H. S. Yanai, W. J. Freund, and O. L. Carter, Thermochim. Acta, 4, 199 (1972).

[129] G. W. Miller, in Ref. 8, p. 435.

[130] G. W. Miller, in Ref. 13, p. 71.

[131] G. W. Miller and J. H. Saunders, J. Appl. Polym. Sci., 13, 1277 (1969); J. Polym. Sci., A-1, 8, 1923 (1970).

[132] G. W. Miller, J. Appl. Polym. Sci., 15, 39 (1971).

[133] D. Machin and C. E. Rogers, Polym. Eng. Sci., 10(5), 300, 305 (1970).

[134] E. A. Colombo and E. H. Immergut, J. Polym. Sci., C, 31, 137 (1970).

[135] S. E. Gordon, in Ref. 8, p. 667.

[136] R. W. Hunter and R. M. Washburn, Amer. Chem. Soc., Div. Org. Coatings Plastics Chem. Preprints, 28(1), 433 (1968).

[137] W. M. Prest, Jr. and R. S. Porter, J. Polym. Sci., A-2, 10, 1639 (1972).

[138] J. K. Gillham, Crit. Rev. Macromol. Sci., 1, 83 (1972).

[139] B. L. Williams, Polym. Eng. Sci., 12(4), 282 (1972).

[140] S. Naganuma, T. Sakurai, Y. Takahashi, and S. Takahashi, Kobunshi Kagaku, 29(2), 105 (1972).

[141] D. A. Seanor, Tech. Methods Polym. Eval., 2, 293 (1970).

[142] R. W. Warfield, in Testing of Polymers, Vol. 1 (J. V. Schmitz, ed.), Wiley (Interscience), New York, 1965, p. 271.

[143] T. J. Gedemer, Soc. Plastic Eng., Tech. Pap., 17, 168 (1971).

[144] N. S. Steck, SPE Trans., 4, 34 (1964).

[145] E. Sacher, Rev. Sci. Instrum., 41, 1885 (1970).

[146] P. Hedvig, J. Polym. Sci., C, 33, 315 (1971).

[147] M. Kisbenyi, Ibid., C, 33, 113 (1971).

[148] P. Hedvig and T. Czvikovszky, Angew. Makromol. Chem., 21, 79 (1972).

[149] W. Wrasidlo, J. Polym. Sci., A-2, 9, 1603 (1971); S. Yalof, W. Wrasidlo, J. Appl. Polym. Sci., 16, 2159 (1972).

[150]   J. Chiu, 4th Annual North America Thermal Analysis Society
        Meeting, Worcester, Massachusetts, June 13, 1973.
[151]   H. P. Vaughan, Thermochim. Acta, 1, 111 (1970).
[152]   E. M. Barrall, II, J. F. Johnson, and R. S. Porter, Appl.
        Polym. Symp., 8, 191 (1969).
[153]   A. J. Kovacs and S. Y. Hobbs, J. Appl. Polym. Sci., 16, 301
        (1972).
[154]   G. W. Miller, in Ref. 14, p. 397.
[155]   R. Kaneko, Kobunshi Kagaku, 26(288), 253 (1969).
[156]   B. Ya. Teitel'baum and N. A. Palikhov, Vysokmol. Soedin., A,
        10(7), 1468 (1968).
[157]   A. R. Shultz and B. M. Gendron, J. Appl. Polym. Sci., 16, 461
        (1972).
[158]   I. C. McNeill, Eur. Polym. J., 3, 409 (1967); 6, 373 (1970).
[159]   F. T. Eggertsen and F. H. Stross, Thermochim. Acta, 1, 451
        (1970).
[160]   J. P. Creedon, Pittsburgh Conference on Analytical Chemistry
        and Applied Spectroscopy, Cleveland, Ohio, March 7, 1972,
        Paper 86.
[161]   D. G. Paul, D. J. Brindle, and J. A. Wegener, Pittsburgh
        Conference on Analytical Chemistry and Applied Spectroscopy,
        Cleveland, Ohio, March 6, 1972, Paper 14.
[162]   F. T. Eggertsen, E. E. Seibert, and F. H. Stross, Anal. Chem.,
        41, 1175 (1969).

Monitoring the Cross-Linking of
Epoxide Resins by Thermoanalytical Techniques

J. M. BARTON

Materials Department
Royal Aircraft Establishment
Farnborough, Hants., England

ABSTRACT

The curing reaction of bisphenol A diglycidylether with 4,4'-
diaminodiphenylmethane (DDM) was studied by thermoanalytical methods. The overall reaction was monitored through
the exothermic heat of reaction by differential scanning
calorimetry (DSC), and a method is developed for predicting
isothermal conversion-time curves over a wide temperature
range from the results of two dynamic DSC scans. The reaction mechanism is not specified but it is assumed not to
change with conversion, and the rate is assumed to be controlled by a single rate constant of the Arrhenius form.
A series of fully cured resins prepared with varying DDM
concentration is characterized by penetrometer, thermal
expansion, and DSC methods. The T_g's of these resins are
compared with those obtained using the stoichiometric
quantity of DDM and reacted to different calorimetric degrees
of cure. The T_g of the resin increases by about $70°C$ in the
final 10% of the curing reaction where ΔH measurements are
least sensitive, so that the final stages of cure are best
monitored by T_g measurements.

25

The value of thermoanalytical techniques for investigating the curing reactions of epoxide resins has been demonstrated previously [1-4]. From a study of the curing reaction of the diglycidyl ether of bisphenol-A (DGEBA) with hexahydrophthalic anhydride, Fava [1] has shown that isothermal cure curves can be predicted from dynamic differential scanning calorimetry (DSC) experiments without any assumptions being made about the form of the kinetic equations or reaction mechanism. A disadvantage of the method is that data from a large number of DSC scans over a wide range of heating rates are required.

In this paper a method is developed for obtaining a profile of resin cure characteristics, and predicting isothermal cure curves, from two dynamic DSC experiments at different heating rates. The method is evaluated for the system DGEBA/4,4'-diaminodiphenylmethane (DDM). The relative sensitivities of different thermoanalytical techniques for assessing extent of cure are also discussed.

It is assumed that the overall rate of cure is given by

$$d\alpha/dt = kf(\alpha) \tag{1}$$

where α is the fractional conversion which for a calorimetric experiment equals $\Delta H_t/\Delta H_0$, ΔH_t being the heat of reaction up to time t and ΔH_0 the overall heat of reaction. For a DSC experiment, therefore, the rate of cure $d\alpha/dt = (dH/dt)(1/\Delta H_0)$. The quantity $f(\alpha)$ is unspecified and depends on the reaction mechanism. The rate constant k is assumed to have the usual Arrhenius form: $k = A \exp(-E/RT)$. If the activation energy, E, is known, then the rate, $r = d\alpha/dt$, determined at one temperature and conversion can be reduced to a corresponding rate at a different temperature and the same conversion:

$$\ln(r_1/r_2) = (E/R)(T_2^{-1} - T_1^{-1}) \tag{2}$$

In this way a rate vs conversion plot obtained from a dynamic DSC experiment can be reduced to corresponding plot at a single temperature. Fava [1] has shown that the area under a plot of $1/r$, for a fixed temperature, against α is the time for a given conversion at that temperature,

$$\int_{\alpha_0}^{\alpha} (dt/d\alpha)d\alpha = t_\alpha - t_{\alpha_0}$$

An isothermal cure curve can thus be built up from the reduced rate vs α data, and using the activation energy this can be transformed into a family of isothermal curves at different temperatures.

E can be obtained from two dynamic DSC scans at different heating rates by the use of Eq. (2) since a given degree of conversion in each of the scans will correspond to different reaction rates and temperatures.

EXPERIMENTAL

The epoxide resin was Shell Epikote 825 which is essentially (96 to 99%) DGEBA. The epoxide equivalent weight is 175. The DDM was recrystallized twice from acetone/toluene. The required amount of DDM was dissolved in the resin using a small quantity of dichloromethane as cosolvent. The dichloromethane was removed under vacuum at 70 to 80°C, and the resin was either used in the DSC curing experiments or was fully cured as a cast sheet (0.5 to 1.5 mm thick) in an oven using the cure schedules given in Table 2.

DSC measurements were made on resin samples of about 20 to 30 mg with the Du Pont 900/DSC instrument, operating in a nitrogen atmosphere. Pure Sn, In, and Hg were the calorimetric standards.

The Du Pont 941 instrument was used for the thermomechanical tests. Any surface oxidized layer on the cast resin samples was removed with fine emery for the penetrometer tests, which were made with a cylindrical quartz probe, 0.025 in. diameter, and a 20-g load. A quartz probe of 0.100 in. diameter was used for the thermal expansion measurements.

RESULTS AND DISCUSSION

The DSC curing experiments were all on the stoichiometric mixture of diepoxide and diamine (28.3 phr of DDM). The total exothermic heat of reaction was found from scans at 20°C/min to be 97.1 cal/g. Data from two DSC scans at heating rates of 10 and 20°C/min are shown in Table 1. The activation energies derived using Eq. (2) at different values of α are also included in the table. Apart from the values for $\alpha = 0.1$ and 0.8, the scatter in E is small. The average value for E, with 95% confidence limits, is 12.6 ± 1.4 kcal/mole, and this is close to the value determined from isothermal experiments: 11.9 kcal/mole.

The data from a DSC scan (10°C/min) were reduced to 100°C using Eq. (2) for E = 12.6 kcal/mole. The original and reduced data are shown in Fig. 1 as plots of r_T (T = 120 to 230°C) and r_{100} against α. The plot of $1/r_{100}$ against α is shown in Fig. 2. Integration of the area under this plot with a lower limit $\alpha_0 = 0.0025$ gave the predicted 100°C isothermal cure curve. These data were transformed into a family of isothermal curves at different temperatures using

TABLE 1. Determination of Activation Energy (E) from DSC Scans at 10 and 20°C/min

α	T_{20} (°C)	T_{10} (°C)	$10^2 r_{20}$ (min^{-1})	$10^2 r_{10}$ (min^{-1})	E (kcal/mole)
0.1	168	150	17.0	8.1	15.29
0.2	179	159.5	28.4	15.7	11.82
0.3	185	165	41.6	23.2	11.66
0.4	189	169	51.0	28.8	11.62
0.5	192.5	172	54.0	30.9	11.23
0.6	196	175.5	51.4	28.7	11.90
0.7	200.5	179.5	42.6	23.1	12.12
0.8	206.5	190	28.4	15.5	16.22
0.9	216.5	194.5	13.6	7.9	11.25

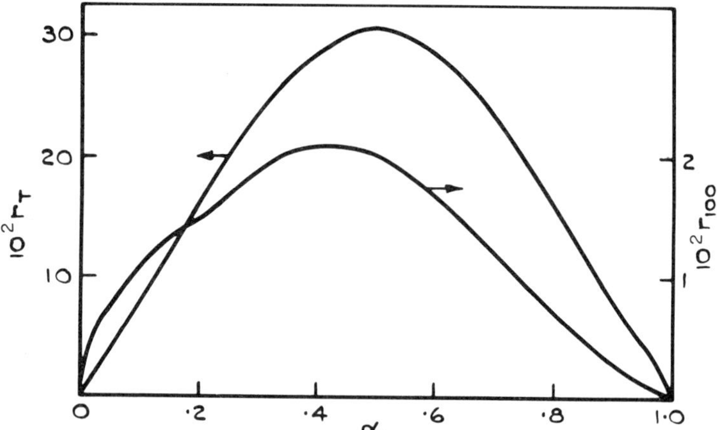

FIG. 1. Dependence of dynamic rate (r_T) and 100°C reduced rate (r_{100}) on conversion for 10°C/min DSC scan.

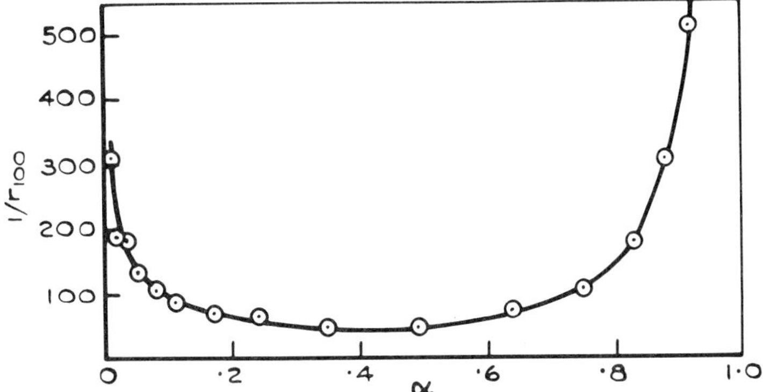

FIG. 2. Dependence of $1/r_{100}$ on conversion.

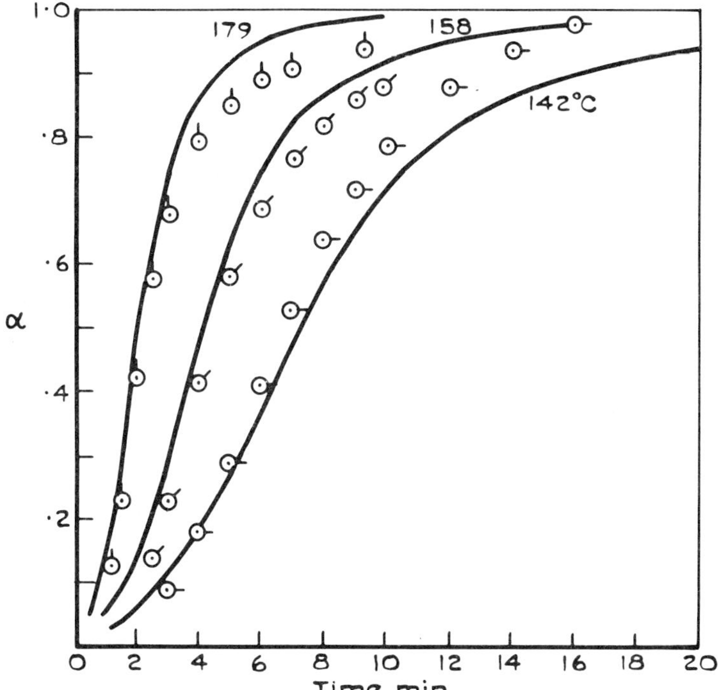

FIG. 3. Isothermal cure curves predicted from dynamic data
(solid curves) and corresponding isothermal experimental points.

Eq. (2) and the relationship $(t_2/t_1)_\alpha = (r_2/r_1)_\alpha$ for temperature T_1 and T_2. The predicted curves are compared with experimental isothermal data in Fig. 3. While exact agreement is not obtained, the predicted curves do provide a fair approximation to those observed. The discrepancies are greater at higher conversions, and this is the region where rate controlling diffusion processes become significant. When the T_g of a cross-linked resin approaches the cure temperature, the curing reaction is retarded and appears to be controlled by segmental relaxation processes [1, 5]. Horie et al. [3] found that the cure of DGEBA by aliphatic diamines became subject to diffusion control at about 60% conversion. These factors are not allowed for in the simple empirical model used, so that discrepancies would be expected at high degrees of conversion. From this standpoint it is perhaps surprising that the activation energy appears not to increase with increasing conversion.

In order to provide standards of comparison for calorimetric degree of cure and fixed degrees of cross-linking, a series of resins was prepared in which the DDM concentration was varied between 74 and 100% of the stoichiometric level. These resins were fully cured and their glass transitions (T_g's) were determined by DSC, penetrometer, and thermal expansion measurements. From the DSC and penetrometer data the inflection points in scans at $20°C/min$ were taken as the T_g. The effect of heating rate on apparent T_g (DSC) was small, a reduction in heating rate from 20 to $50°C/min$ caused a reduction in T_g of only $3.5°C$ for the 100% stoichiometric sample. In the thermal expansion measurements the samples were equilibriated at about $T_g + 50°C$ and the contraction was then observed at a cooling rate of $2°C/min$. The T_g was obtained as the intersection of the approximately linear contraction curves extrapolated from above and below T_g. The results are shown in Table 2. The penetrometer T_g's are 2 to $20°C$ lower than those determined by DSC, with the greatest differences arising in the most highly cross-linked samples. This can be ascribed to the known stress activation of relaxation processes in polymers. It has been observed that heating experiments tend to show higher T_g's than those using a cooling mode because of nonequilibrium in the molecular relaxation processes [6], and this is shown in the lower T_g's obtained from contraction during cooling from above T_g compared to the DSC specific heat transitions obtained by heating from below T_g. A monotonic decrease in apparent T_g with decreasing degree of cross-linking is observed with all three methods used, and each would therefore be useful

TABLE 2. Effect of DDM Concentration on Resin Properties

DDM	Cure time (hr)		T_g ($^\circ$C)		
% stoichiometric	150°C	200°C	DSC	Penetrometer	Expansion
99.9	4	3	184	172	163
94.9	20	2	174	154.5	158
90.1	20	2	166	160.5	160
87.5	20	-	153	146.5	140
80.3	20	-	132	125.5	120
74.2	20	-	119.5	117.5	113

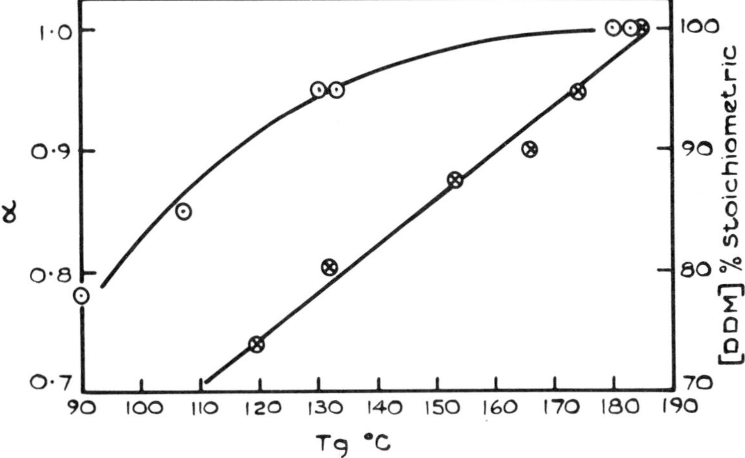

FIG. 4. Dependence of T_g on conversion and DDM concentration.

for monitoring the relative degree of cross-linking in the present system.
For routine work the penetrometer test would be very useful as it is experimentally simple and generally yields the most clearly defined
transitions. The approximately linear dependence of T_g on DDM concentration is shown in Fig. 4.

From isothermal experiments at various cure temperatures and times,
samples were obtained of known calorimetric fractional conversion, α.
The T_g's of these samples determined by DSC are also shown in Fig. 4.
It is seen that T_g is a very sensitive index of the degree of cure and that

there is a large increase of about $70°C$ in T_g during the final 10% of the curing reaction where ΔH measurements are least sensitive. Comparison with the plot of T_g of the fully cured samples against initial DDM concentration shows that the T_g for a given fractional conversion is lower than that corresponding to the same fractional stoichiometric quantity of DDM except in the last few percent of conversion. These differences may be ascribed to the plasticizing effect of unreacted diamine in the undercured stoichiometric compositions.

In conclusion, the early part of the curing reaction may be quantitatively described to a fair approximation using data from only two DSC scans. This method should be useful for other systems where Eq. (2) applies, k is of the Arrhenius form, and the reaction mechanism does not change during the reaction. Large increases in T_g occur during the final 10% of the curing reaction, and this phase is best monitored by T_g measurements.

REFERENCES

[1] R. A. Fava, Polymer, 9, 137 (1968).
[2] O. R. Abolafia, SPE Ann. Tech. Conf., Preprints, 15, 610 (1969).
[3] K. Horie et al., J. Polym. Sci., A-1, 8, 1357 (1970).
[4] M. A. Acitelli et al., Polymer, 12(5), 335 (1971).
[5] M. Gordon and W. Simpson, Ibid., 2, 383 (1961).
[6] S. M. Wolpert et al., J. Polym. Sci., A-2, 9, 1887 (1971).

Evaluation of Initiators and Fillers in Diallyl Phthalate Resins by Differential Scanning Calorimetry

PAUL E. WILLARD

FMC Corporation
Princeton, New Jersey 08540

ABSTRACT

Seven organic peroxide initiators for the polymerization of diallyl-o-phthalate prepolymers were studied using differential scanning calorimetry. Included were t-butyl perbenzoate, dicumyl peroxide, α,α'-bis(t-butyl peroxy) diisopropyl-benzene, 2,5-dimethyl-2,5-di(t-butyl peroxy) hexane, 2,5-dimethyl-2,5-di(t-butyl peroxy)hexyne-3, di-t-butyl peroxide, and t-butyl hydroperoxide. Heats of reaction and reaction rate constants are presented for each initiator at 1, 2, 3, and 4 phr of diallyl-o-phthalate prepolymers. The differences in reaction temperature and rate are discussed. Effects of four types of commonly used fillers (asbestos floats, ground quartz, calcium silicate, and clay) on the heats of reaction of diallyl-o-phthalate prepolymers using t-butyl perbenzoate and dicumyl peroxide initiators show the large inhibiting effect of untreated kaolinite clays on this polymerization.

33

INTRODUCTION

Doehnert and Mageli [1] measured the decomposition of organic
peroxide initiators in benzene and acetone solutions. Their method
allowed the calculation of activation energies and rate constants for
the decomposition of these initiators in dilute solutions. The
technique had not been extended to studying the actual decomposition
kinetics of these initiators in polymeric materials, although such
work was proposed in their paper.

Barrett [2] studied the thermal decomposition of benzoyl peroxide
and azobisisobutyronitrile in solutions of di-n-butyl phthalate. His
method, based on differential scanning calorimetry (DSC), allowed
the calculation of kinetic constants for thermal decomposition from
a single DSC curve.

The present work involves the study of seven commercial peroxide initiators in diallyl phthalate prepolymer. The seven initiators
studied were:

t-butyl perbenzoate
Di-Cup (registered trademark, Hercules, Inc.) (dicumyl peroxide)
Vul-Cup (registered trademark, Hercules, Inc.) (α,α'-bis(t-butyl
peroxy)diisopropyl-benzene)
Lupersol 101 (registered trademark, Lucidol Division, Pennwalt
Corp.) (2,5-dimethyl-2,5-di(t-butyl peroxy)hexane)
Lupersol 130 (registered trademark, Lucidol Division, Pennwalt
Corp.) (2,5-dimethyl-2,5-di(t-butyl peroxy)hexyne-3)
di-t-butyl peroxide
t-butyl hydroperoxide

Table 1 summarizes the properties of the catalysts used in this
study and gives 1 min half-life data in dilute solution in benzene.

These seven initiators were studied at four different concentrations in Dapon 35 (registered trademark, FMC Corp.), a prepolymer
of diallyl-o-phthalate.

The data presented herein are in terms of initiator (or catalyst,
as it is known in the industry) as parts per hundred of resin (phr).
This convention is widely used in diallyl phthalate compounding.

The effect of four commonly used fillers was also determined
with both t-butyl perbenzoate and Di-Cup initiators.

EXPERIMENTAL

DSC curves were taken with a Du Pont Model 900 Thermal
Analyzer with a DSC cell. This instrument was interfaced to an
analog to digital converter and punched paper tape generated as the
scan was carried out as described previously [3].

TABLE 1. Catalyst Properties

Material	Abbreviation used in this report	% Active oxygen	1 Min half-life at T (°C)	1 Min half-life at T (°F)
t-Butyl perbenzoate	TBP	8.1	166	331
Di-Cup	Di-Cup	11.9	174	345
Vul-Cup	Vul-Cup	18.9	180	356
Lupersol 101	L-101	10.5	174	346
Lupersol 130	L-130	5.6	193	379
Di-t-butyl peroxide	DTBP	10.8	193	380
t-Butyl hydroperoxide	TBHP	12.7	179	354

Using the method of Barrett, activation energies and rate constants were determined for each of the concentrations of initiators.

All samples were prepared by ball milling the initiators into the powdered prepolymer. Replication of tests showed the homogeneity of the mixtures. All DSC curves were taken in a nitrogen atmosphere at a heating rate of 10°C/min. Sample weights were approximately 30 mg and consisted of preforms made by cold pressing the prepolymer-initiator mixtures at pressures of 100,000 psi at room temperature in a 1/4-in. pellet press die. This procedure insured that all samples had the same density and surface volume relationships regardless of particle size. This technique is also applicable to the study of molding compounds where granular size can influence the results obtained by DSC.

RESULTS AND DISCUSSION

Figure 1 presents the DSC curves for seven commercial peroxide initiators at a concentration of 1 pph of Dapon 35 prepolymer. There is a wide range in temperature activity of these peroxides, ranging from 169°C for t-butyl perbenzoate up to as high as 224°C for t-butyl hydroperoxide. The peak temperatures indicated on the figure are

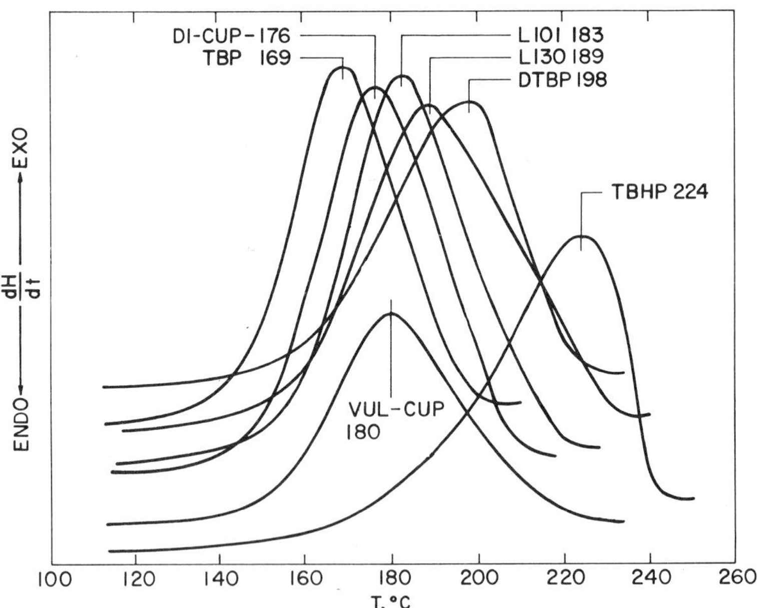

FIG. 1. Effect of catalyst type.

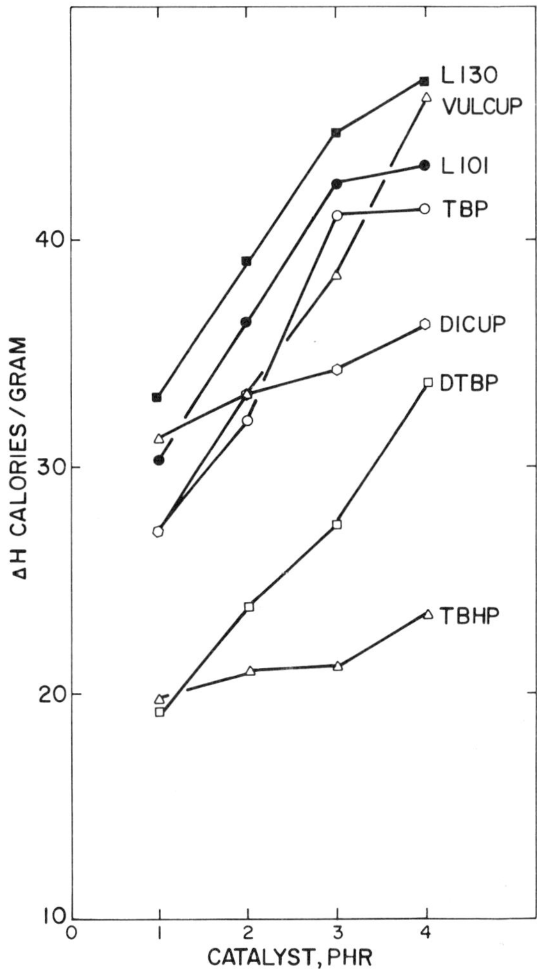

FIG. 2. Catalyst type and level vs heat of reaction.

the peaks in the DSC curve, representing the temperature at which the reaction is proceeding at its maximum rate. These figures can be used to compare the activity of the various peroxides in the diallyl phthalate prepolymer.

The data from the DSC curves, as punched paper tapes, were analyzed by computer program [3] and the total heat of reaction (area under the DSC curve) was determined as a function of concentration for each initiator.

Figure 2 shows the data obtained and indicates the steep concentration dependence of these initiators. Note that for Lupersol 130 at 4 parts of initiator phr, the thermal output is approximately 47 cal/g. At 1 phr, the thermal output is only 33 cal/g. Similar concentration dependencies are shown for the other initiators. This finding is important to the manufacturer of commercial molding compounds since lowering catalyst concentration may give very low degrees of cure in the compound.

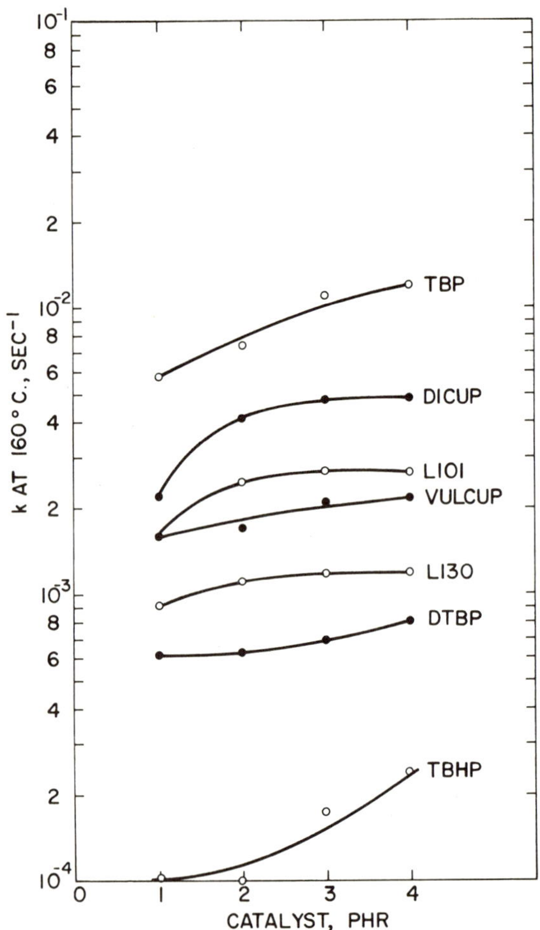

FIG. 3. Catalyst type and level vs reaction rate constant.

There is also a wide difference in the total heat of reaction obtainable with the various initiators, and this initiator efficiency should be known if such catalyst systems are to be used in commercial compounds. Thus t-butyl perbenzoate (TBP), which is a commonly used commercial material, is much more efficient than t-butyl hydroperoxide or di-t-butyl peroxide.

Reaction rate constants were calculated, and Fig. 3 presents the reaction rate constants at 160°C in $\sec^{-1}$ as a function of concentration. With the exception of t-butyl hydroperoxide (TBPH), there is not as large a concentration dependency in the case of reaction rate constant k as there is in the case of heat of reaction ΔH. However, since these data in Fig. 3 are presented at 160°C, a commonly used molding temperature for diallyl phthalate, the activity of the initiators covers almost two decades of reaction rate constant. Thus the compounder who uses t-butyl hydroperoxide at conventional molding temperatures will find the reaction proceeding almost 100 times slower than that of TBP.

Solomon [4, 5] has found that certain kaolinite clays seriously degrade the performance of peroxide initiators in polymerization reactions. He found that kaolinite clays were particularly effective in decomposing organic peroxide initiators because of the high acidity of the clay surface.

In order to test these effects in diallyl phthalate prepolymers, mixtures of four common fillers at the 50% level were made with Dapon 35 using two different organic peroxide initiators. Four materials used were: asbestos 7 TF-1 floats (Johns-Manville), silica 219 (ground quartz) (Harshaw Chemical Co.), Wollastonite T-1 (natural calcium silicate) (Cabot Corp.), and an untreated kaolinite clay (ASP 400) (Georgia Kaolin Co.). These materials were added at equal weights to Dapon 35 catalyzed with 3 phr of t-butyl perbenzoate giving a 50% by weight filler loading. DSC curves were then run on these compositions. As shown in Fig. 4, total heat of reaction was least affected by the ground quartz (silica 219) which gave a ΔH value of 20.6 cal/g, which is exactly half the value obtained for Dapon 35 and 3 parts t-butyl perbenzoate with no filler. Wollastonite reduced the total heat reaction slightly, as did asbestos. The clay reduced the value to 13.1 cal/g or 64% of the value obtained with silica 219.

When Di-Cup (dicumyl peroxide) initiator was used in the same 50% filler loading with these fillers, the effect of the untreated kaolinite clay, ASP 400, was extremely high. Figure 5 shows the DSC curves using Di-Cup initiator. Here the silica value is lower than that obtained with t-butyl perbenzoate (which we would expect from the relative reactivities of TBP and Di-Cup). However, in the case of ASP 400 clay, total heat of reaction drops to 0.3 cal/g which is 1.6% of the value obtained with silica 219 in this initiator system.

Obviously, the use of ASP 400 clay in Di-Cup initiated systems may result in materials which will not cure due to decomposition of the initiator by the acidic clay surface.

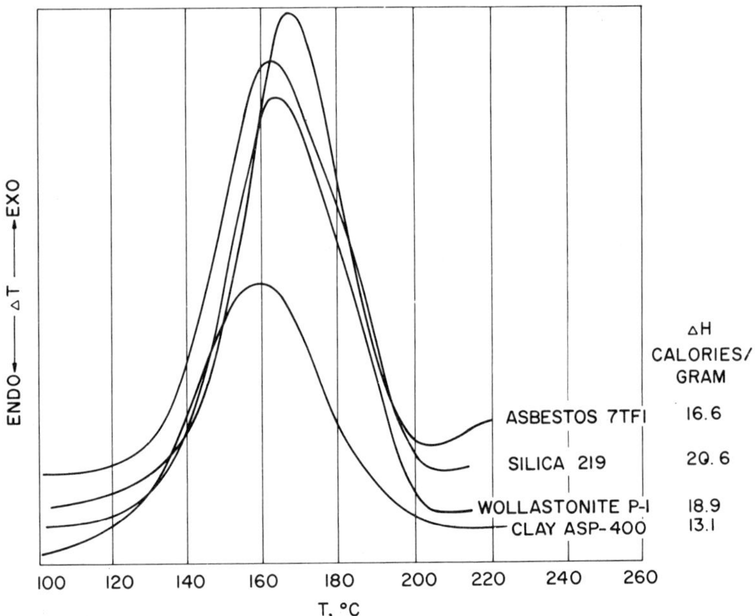

FIG. 4. Effect of fillers on cure–TBP.

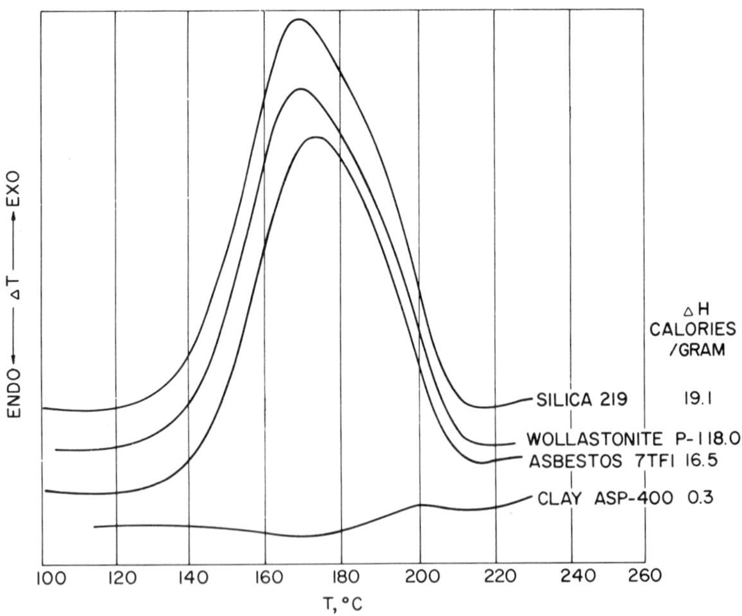

FIG. 5. Effect of fillers on cure—Di-Cup.

The use of DSC to measure activity of initiators for the diallyl phthalate system and to determine the effects of fillers on total heat of reaction allows rapid screening of initiator systems and compounding ingredients to determine their activities in the polymerization. By using small quantities of materials to obtain DSC curves, it is possible to survey large numbers of compositions which would be impossible to study using conventional compounding molding and physical testing techniques. DSC allows the rapid and inexpensive screening of these systems to allow predictions of polymerization performance.

ACKNOWLEDGMENTS

The author thanks FMC Corp. for permission to publish this material. B. K. McEuen performed most of the DSC experiments.

REFERENCES

[1] D. F. Doehnert and O. L. Mageli, Mod. Plastics, 36, 142 (1959).
[2] K. E. J. Barrett, J. Appl. Polym. Sci., 11, 1617 (1967).
[3] P. E. Willard, SPE RETEC, "The Challenge of Thermosets,"
     Chicago, February 1973.
[4] D. H. Solomon et al., J. Macromol. Sci.—Chem., A5(3), 587
     (1971).
[5] D. H. Solomon et al., Ibid., A5(5), 995 (1971).

Novel Melting Phenomena in a Series of Isothermally Crystallized Polyethylene Single Crystals

IAN R. HARRISON

Department of Material Sciences
The Pennsylvania State University
University Park, Pennsylvania 16802

ABSTRACT

Polymer single crystals or lamellae are widely used in
models for more complex systems such as fibers, films,
or bulk polymer. Determination of important thermodynamic and physical properties for the crystals is often
frustrated by their tendency to reorganize on heating.
Partly as a consequence of this, the melting ranges for
high molecular weight polymers are often quite broad. A
novel technique has been developed which permits a more
detailed examination of the melting/morphology relationships that exist in single crystals. This has led to the
observation that high molecular weight polyethylene (PE)
single crystals possess much narrower melting ranges than
had been previously reported. It has been possible to
determine distinct transformation temperatures for the
$\{100\}$ and $\{110\}$ fold sectors that exist in some PE crystals.
Melting point determinations have been made on a series
of isothermally crystallized PE crystals using DTA.
Narrow multiple peaked thermograms were obtained, and
peak position and ratio could be related to morphological
differences between the crystals.

43

INTRODUCTION

Polymer single crystals, or rather polymer lamellae, are widely
used in developing models for fibers, films, and bulk polymer. To
better understand the behavior of these commercially important
classes of materials, it is necessary to study the properties of
lamellae or single crystals. Measurement of the melting point, dissolution temperature, or heat of fusion of such crystals allows the
deduction of important thermodynamic and physical properties [1].
However, these crystals tend to anneal [2], recrystallize [3], or
superheat [4] during the heating cycle. This prevents accurate
determination of the above parameters. Efforts have been made to
allow for these reorganizational processes with only limited success.
Most thermograms for high molecular weight polymers are
characterized by their breadth relative to those for low molecular
weight organic solids. Besides the reorganizational processes
previously mentioned, this breadth is also attributed to a large distribution of crystallite sizes [5] or to degrees of perfection within
the crystal.

A technique has been developed which demonstrates that broad
thermograms can be resolved into a number of narrow melting
peaks. Further, this technique permits the electron microscopic
examination of single crystals whose thermal history has been recorded in the Differential Thermal Analyzer. This technique was
applied to a series of polyethylene (PE) single crystals grown under
isothermal crystallization conditions. Results of this study, which
produced novel information regarding the melting process, are
presented in this paper.

EXPERIMENTAL

Details of the crystallization procedure are given elsewhere [6].
However, it should be noted that crystals were grown from Marlex
6001, a high molecular weight linear PE. Crystallization was performed isothermally at a number of different temperatures. The
crystals were all washed three times by a decantation process at
their particular crystallization temperature (T_c). This washing
reduced the concentration of impurities, primarily of low molecular
weight PE, to less than one-thousandth of their original concentration.

After washing, the crystals were allowed to come to room temperature. Part of each batch was exchanged to acetone by repeated
centrifugation and washing with acetone. Some of the acetone exchanged sample was allowed to air dry and was finally dried under
vacuum for several days. A portion of the acetone suspension was
also exchanged to silicone oil. This exchange was achieved by adding

oil to the suspension in acetone (or xylene) and then evaporating the original suspending medium. Complete evaporation of acetone or xylene was obtained by application of a high vacuum for several days. All exchanges, evaporations, etc., were carried out at room temperature.

Differential scanning calorimetry (DSC) and differential thermal analysis (DTA) runs were made on both the dried down sample (solid) and the oil exchanged suspensions (exch. to oil). These runs were made on various sample weights at different heating rates and sensitivities. However, this work will be primarily concerned with DTA runs on "exch. to oil" samples run at 20°C/min. All DTA runs were carried out in 2 mm tubes using a DuPont 990 Thermal Analyzer and Cell Base with a standard DTA cell. All runs reported have been at least duplicated and often five or more scans were made.

RESULTS AND DISCUSSION

Sample sizes for DSC and DTA are typically on the order of 1 to 6 mg [1-5, 7, 8]. Thermograms obtained from this size sample of high molecular weight PE are usually broad, simple fusion curves, covering some 10 to 15°C [1]. Lower molecular weight samples give multiple peaks, generally interpreted as evidence of some reorganizational process [3]. Radiation-induced cross-linking has been used to inhibit reorganization and produces narrower fusion curves [2, 8]. However, irradiation does not simply cross-link the crystal surfaces [7], leading one to question the significance of melting temperatures so obtained.

A study was made of the effect of "solid" sample size on the thermogram. This led to a study of "exch. to oil" crystals and the effect of heating rate on their thermograms. This work has been covered in more detail elsewhere [6]; however, for completeness a summary is included here. Finally the effect of T_c on the thermogram was examined.

DTA thermograms were obtained on various weights of the "solid" sample for one crystallization temperature, namely 86°C. Two of these runs at 20°C/min are shown in Fig. 1. At the higher weight, 2.76 mg, a broad simple fusion curve was obtained. As sample weight

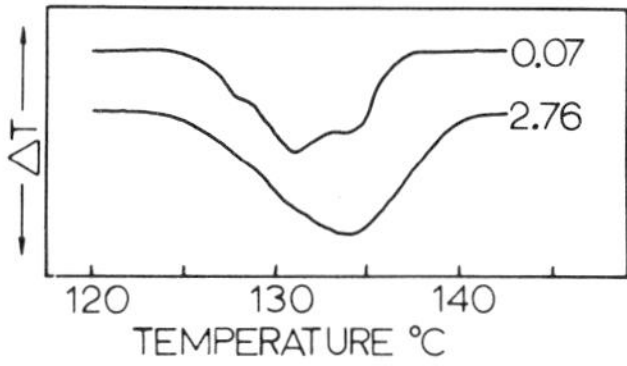

FIG. 1. DTA runs on "solid" sample at 20°C/min. Weight of sample given in milligrams.

decreased the thermogram became more complex and eventually at
the lower weight, 0.07 mg, three peaks were discernable. These are
contained within the spread of the broad peak from the larger sample.

After melting, these samples were recrystallized from the melt by
cooling at a programmed rate. When thermograms were run on these
melt-recrystallized materials, a single broad peak was obtained. The
shape and position were essentially the same, regardless of sample
weight. This observation suggests that sample weight differences
shown in Fig. 1 may be due to poor thermal conductivity of the
powdered sample. As the weight of a powder sample is increased,
the distribution of thermal states becomes increasingly broader. The
thermocouple detects the average of this distribution of states, resulting in a single broad peak at higher weights.

In an effort to check this idea, the "solid" sample was dispersed in
silicone oil. This is a nonsolvent for the polymer and has the same
order of magnitude thermal conductivity as solid polymer. Thermograms from these dispersions indicated that for a given weight of
polymer the oil-dispersed samples had narrower multiple peaks than
the "solid." This was true up to the highest sample weight used.

The narrowest multiple peaked thermograms were obtained when
the crystals were dispersed in oil without ever drying them down,
i.e., "exch. to oil" samples. A comparison between the thermograms
of "solid" and "exch. to oil" samples for the same crystal preparation
and with approximately the same weight of polymer is shown in Fig. 2.
Note the narrowness of the "exch. to oil" peaks, approximately $1.5°$C
at half peak height for the largest peak in the thermogram. Peaks this
narrow have not previously been reported for any polymer sample. It
should also be noted from Fig. 3 that the positions of the peak maxima
for the "exch. to oil" samples are independent of heating rate in the
range 5 to $80°$C/min. Peak position is unchanged but the thermogram
width seems to increase as heating rate is increased.

The "exch. to oil" thermogram shown in Fig. 2 is apparently
composed of three peaks; the two low temperature peaks are narrow,
the high temperature peak is somewhat broader. These peaks will be

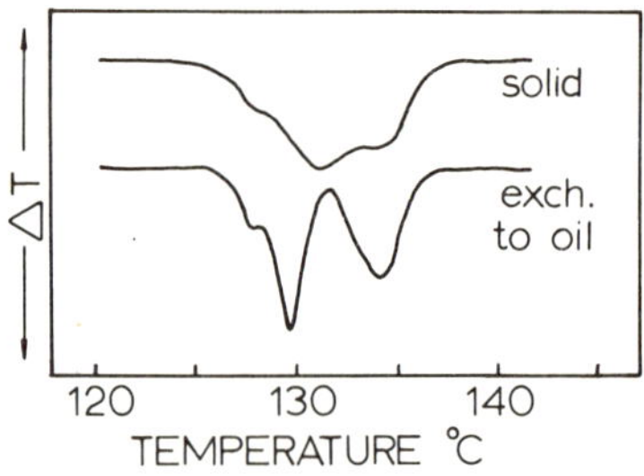

FIG. 2. DTA runs on "solid" and "exch. to oil" samples. Samples
both have approximately 0.1 mg of polymer and were run at $20°$C/min.

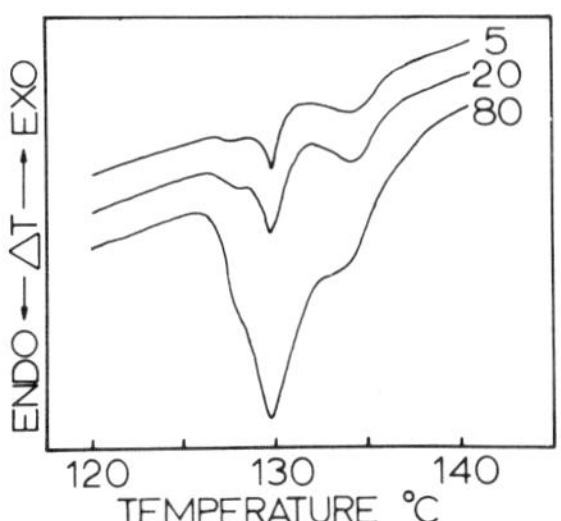

FIG. 3. DTA runs on "exch. to oil" samples of approximately 0.1 mg. Heating rate is given in °C/min.

referred to as first, second, etc., in order of increasing temperature of their endotherm maximum. A sample was heated just to the maximum of the first peak, quenched, then reheated. The resulting thermogram is shown in Fig. 4(a). The first peak has completely disappeared, the second peak remains unchanged, and the third peak shows some slight change in shape.

With regard to the origin of the peaks, in light of previous work on annealing and recrystallization, it is suggested that the third peak represents the melting behavior of crystals which have already under-gone some reorganization. Additional evidence will be presented later. The presence of a first and a second peak could be the result of im-purities including a separate crystalline phase, or a phase transition in-cluding annealing and/or recrystallization. However, no rate dependence typical of phase transitions [9] was observed (Fig. 3). In addition, previous studies have found no evidence for separate crystalline phases [10]. The care taken to crystallize isothermally and wash the samples tends to rule out the presence of impurities, certainly in the quantities

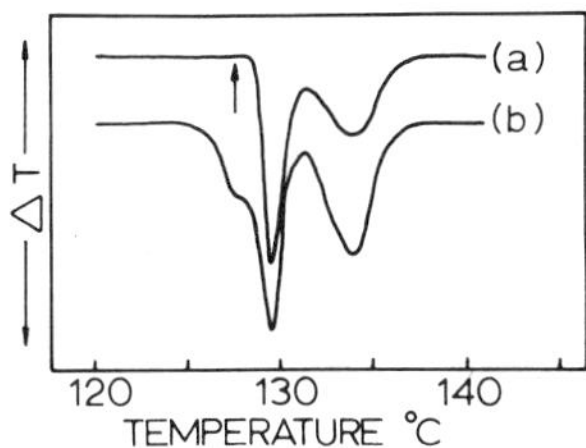

FIG. 4. (a) An "exch. to oil" sample was heated just to the point indicated by the arrow and quenched. On reheating the thermogram was obtained. The point indicated by the arrow corresponds to the first peak maximum. (b) For comparison, the thermogram of an untreated "exch. to oil" sample.

indicated by the peak size. The implication is therefore that isolated lamellae of PE crystallized at 86°C exhibit two independent melting transitions.

As widely known, PE crystals grown at about this temperature from xylene exist as truncated lozenges with $\{100\}$ and $\{110\}$ fold sectors [10]. It has been suggested that a definite difference in transformation temperature should exist for these two sectors [11]. Studies performed to date, with DSC have not revealed such a difference. It is suggested that the first and second peaks discussed above represent the transformation temperatures of the two sectors. Previous annealing studies and heat treatments have been carried out on lamellae sedimented on carbon grids [11-13]. In line with these experiments, crystals heated to the first peak maximum in the DTA should show evidence of preferential reorganization of one of the sectors, if the model is correct. An electron micrograph from a sample of those crystals which gave Fig. 4(a) on reheating is shown in Fig. 5. The following are observed on this and other micrographs: darkening of all or part of the $\{100\}$ sectors; the ends of the $\{100\}$ sectors are bowed toward the crystal center; and striations develop which are parallel to $\{010\}$ planes in $\{100\}$ sectors.

The above observations are taken as evidence for preferential melting and subsequent recrystallization of $\{100\}$ sectors. These observations were made only when crystals were heated to the first peak maximum. Crystals heating just 1.5° lower, to the onset of melting, show no such changes. Crystals were also heated to the second peak maximum, quenched, and examined under the electron microscope. Micrographs are not shown; however, the crystals appeared as disk-shaped objects with no structural detail being observed on their surfaces. This is taken as evidence that the $\{110\}$ sectors have now also melted and recrystallized on quenching. It is concluded that the first and second peaks as shown in Fig. 4(b) represent the melting of $\{100\}$ and $\{110\}$ sectors, respectively.

One should also note that the observation of two melting peaks for these crystals implies the existence of different surface energies for the two fold sectors. This is based on the assumption that the lamella thickness is the same for the two sectors. The existence of two surface energies suggests that there may exist different chemical reactivities for the two sectors. Different reactivities have not so far been reported. However, previous chemical studies have only been carried out on dried down crystals. These may have undergone some surface damage during the drying process.

DTA measurements have also been carried out on single crystals grown isothermally at a number of different temperatures. These measurements were made at various heating rates on "exch. to oil" samples. Figure 6 shows typical scans made at 10°C/min. The following are observed: the ratio of the first two peaks to the third peak gradually increases as T_c increases; where three peaks exist,

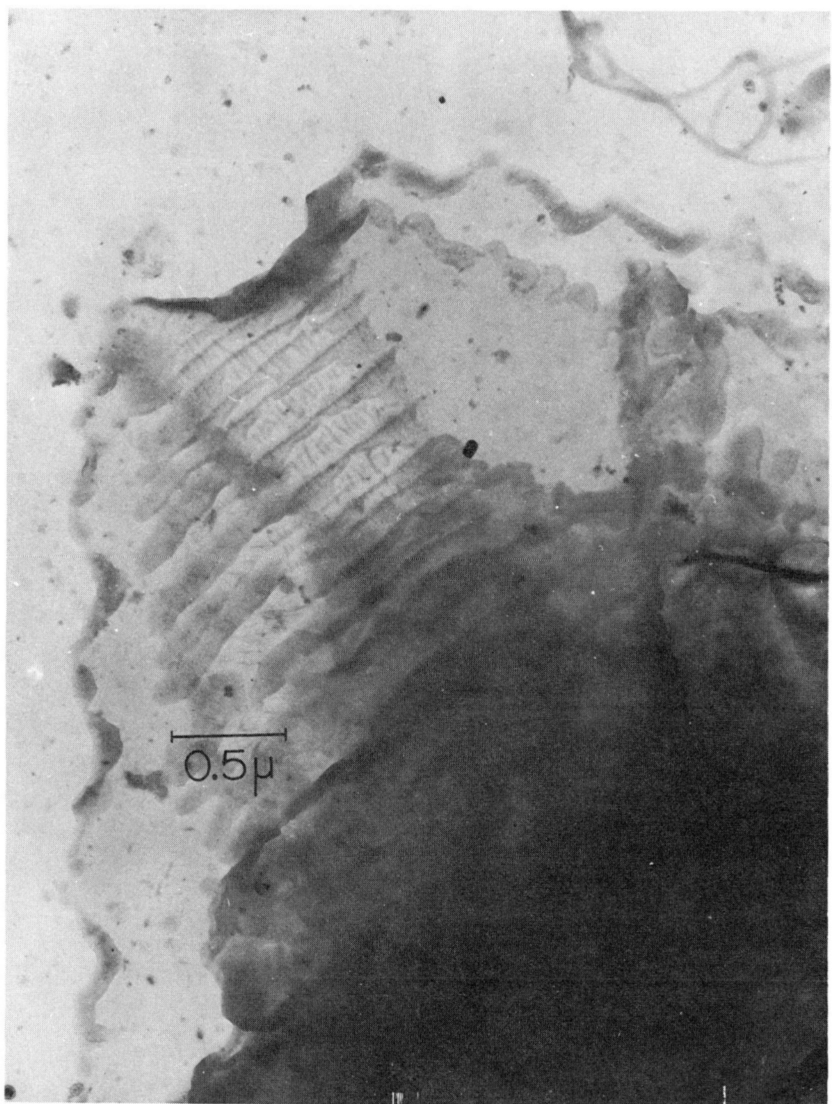

FIG. 5. An electron micrograph of a sample of crystals that gave Fig. 4(a) on reheating. The bowed end and {010} striations in the {100} sector are clearly visible.

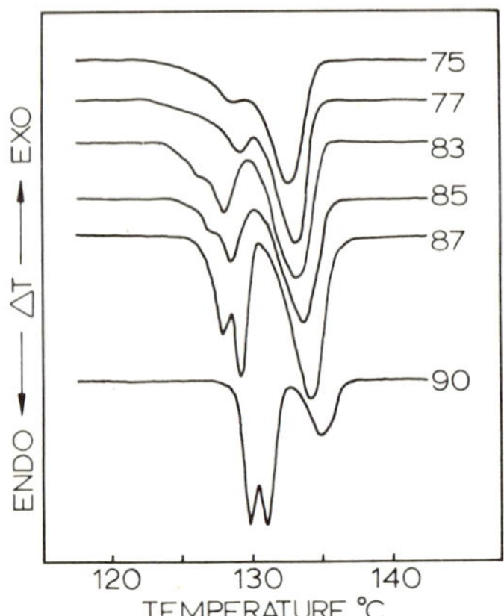

FIG. 6. A series of DTA curves obtained on "exch. to oil" samples.
The isothermal crystallization temperature of the original crystals is
given for each curve in °C. Sample weights were within the range 0.07
to 0.15 mg, and all runs were made at 10°C/min.

the ratio of the first peak to the second peak increases with increasing
T_c; and with the exception of the first two crystallization temperatures,
all peak positions move to higher temperatures with increasing T_c.
Peak positions (T_m) from DTA runs at 20°C/min are plotted as a
function of T_c in Fig. 7. Similar observations may be made for DTA
scans at both rates. However, additional data can be derived from the
plots made in Fig. 7.

For example, the curves for the first and second peaks appear to
intersect, if extrapolated, at a T_c of approximately 96°C. Similarly,
curves for the first and third peaks intersect at a T_c of 103°C and a
T_m of 139.5°C, and curves for the second and third peaks intersect at
a T_c of 110°C and a T_m of 141°C. Additional data and more sophisticated analysis of the thermograms are necessary before much significance can be placed on these particular points of intersection.

Data from 75 and 77°C crystals appear to be out of line with that
derived from crystals grown at higher temperatures. However, two
points should be remembered: 1) crystals grown at 75 and 77°C do
not possess distinct {100} fold sectors, and 2) the lower the crystallization temperature, the more easily will crystals reorganize. In
fact, the <u>onset</u> of melting of all the crystals falls on a straight line as

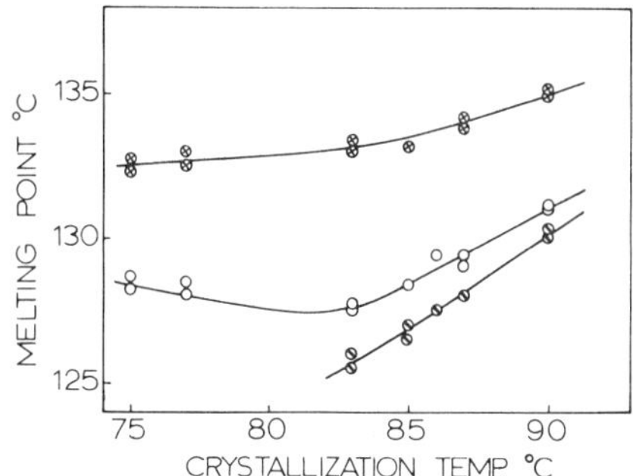

FIG. 7. The two or three melting points (peak maxima) were recorded for a series of isothermally crystallized PE samples. All DTA runs were made at 20°C/min. ⊘ , ○ , and ⊗ represent first, second, and third melting peak respectively, in order of increasing temperature of peak maxima.

a function of T_c when run at 20°C/min. Changes observed in the ratio of the first and second peaks to the third and in the ratio of first to second are consistent with our knowledge of the morphological changes which take place as T_c increases. For example, as T_c increases, lamella thickness increases; there would be higher melting points for the crystals and correspondingly lower probability of reorganization during heating. Similarly, as T_c increases a larger fraction of the polymer exists in the {100} fold sectors [14]. This would lead to the observed changes in ratio of first to second peaks.

CONCLUSIONS

1. A technique has been developed which demonstrates that high molecular weight PE, in single crystal form, possesses much narrower melting ranges than had previously been reported. This observation should aid in the elucidation of polymer crystal structure.

2. Using the above technique it has been possible to demonstrate that {100} and {110} fold sectors in truncated PE crystals possess different transformation temperatures. This leads to the suggestion that these sectors should possess different chemical reactivities.

3. The above observation was made on crystals grown at a single crystallization temperature. Data derived from a series of isothermally crystallized PE samples appear to support this observation. Work is being continued on this series of crystals: (a) to develop a

more sophisticated analysis of the thermograms, and (b) to examine the possibility of selective sector reactivity.

ACKNOWLEDGMENTS

This work was primarily supported by an NSF Research Initiation Grant. The Sandia Laboratories, Albuquerque, New Mexico, permitted the author to continue part of this work, with their support, during the summer of 1972. Microscopy was performed by C. R. Hills and W. W. Corbett of Sandia and Pennsylvania State University, respectively. G. L. Stutzman of Pennsylvania State University performed many of the DTA runs. The author would like to thank all of the above for their kind support and interest.

REFERENCES

[1] For example, see L. Mandelkern, Progr. Polym. Sci., 2, 165 (1970).
[2] H. E. Bair, R. Salovey, and T. W. Huseby, Polymer, 8, 9 (1967).
[3] L. Mandelkern and A. L. Allou, Jr., J. Polym. Sci., B, 4, 447 (1966).
[4] E. Hellmuth and B. Wunderlich, J. Appl. Phys., 36, 3039 (1969).
[5] B. Ke, J. Polym. Sci., 42, 15 (1960).
[6] I. R. Harrison, J. Polym. Sci., Polym. Phys. Ed., 11, 991 (1973).
[7] A. M. Rijke and L. Mandelkern, Ibid., B, 7, 651 (1969).
[8] H. E. Bair and R. Salovey, J. Macromol. Sci—Phys., B3, 3 (1969).
[9] J. L. Koenig and A. J. Carrano, Polymer, 9, 401 (1968).
[10] For example, see P. H. Geil, Polymer Single Crystals, Wiley (Interscience), New York, 1963.
[11] H. Nagai and N. Kajikawa, Polymer, 9, 177 (1968).
[12] D. C. Bassett, F. C. Frank, and A. Keller, Nature, 184, 810 (1959).
[13] A. Keller and D. C. Bassett, J. Roy. Microscop. Soc., 79, 243 (1960).
[14] D. J. Blundell and A. Keller, J. Macromol. Sci.—Phys., B2(2), 337 (1968).

Characterization of Thermosetting Resins by Thermal Analysis

T. R. MANLEY

Department of Materials Science
Newcastle upon Tyne Polytechnic
Newcastle upon Tyne, England

ABSTRACT

Thermoanalytical techniques are valuable tools for the characterization of thermosetting polymers. Examples are given of the application of thermal methods to the identification of filled and unfilled epoxy and amino resins. Various thermoanalytical methods of following the curing of thermosetting polymers are considered. Thermal techniques are essential to the study of degradation reactions; the decomposition of a melamine formaldehyde resin is discussed in detail. It is concluded that the usual methods of thermal analysis alone are insufficient to characterize fully the degradation of amino resins.

Thermosetting resins are intractable materials. They are frequently filled or reinforced with inorganic materials which inhibit normal spectroscopic techniques. While thermal analysis is no panacea, it can prevent headaches in those who are required to identify resins, to find the degree of cure, and to study degradation reactions.

53

IDENTIFICATION OF POLYMERS

Epoxy Resins

Initial epoxy resins were based on bisphenol (Shell Epikote 828, Ciba-Geigy CT200) but cycloaliphatic materials (e.g., CY175) and mixtures (X83/260) are now encountered. Figure 1 shows that the differential thermal analysis (DTA) curves of common commercial resins are sufficiently distinctive for them to be distinguished one from another. Thermogravimetry (TG) may also be used to distinguish between these reactions [1].

When fillers are to be identified, the advantages of thermal analysis are most apparent. Hydrated alumina $Al(OH)_3$ is widely used because of the improvement in tracking properties, and calcium carbonate is employed where castings have to be machined. Resins filled with these materials can be readily distinguished by TG (Fig. 2) or by DTA. Other common fillers, such as glass silica and asbestos, can also be identified by thermal analysis [2].

Thermogravimetry is a convenient method for the quality control of epoxy resins. Figure 3a shows Epikote 828 containing 45% of phthalic anhydride, and Fig. 3b, 15%. With excess hardener the initial loss of weight is due to sublimation of uncombined anhydride. The DTG next shows, beginning at $220°C$, the breakdown of the ester links with the liberation of carbon dioxide. A small peak at $360°C$ is not well resolved due to the large amount of material lost in the decarboxylation process. The weight losses beyond $400°C$ are attributed to oxidation reactions.

In the case of the resin cured with insufficient hardener, there is no free anhydride to sublime and the decomposition begins just below $200°C$. This figure gives a better indication of the true decarboxylation temperature.

Amino Resins

Characteristic DTA curves are obtained for melamine and urea formaldehyde resins. These enable one to distinguish between cured and uncured specimens and also to say whether combinations of melamine with urea or phenol are mixtures or genuine copolymers.

A cured melamine formaldehyde (MF) resin contains methylol groups, ether groups, and methylene groups. The proportions of these three groups per triazine ring varies but all three groups may be found in normal cured molding powder.

When heated in a closed cell (Fig. 4a), the double endotherms of curing are clearly seen followed by the characteristic exotherm near $360°C$. Figure 4b shows the DTA curve of resin cured under normal

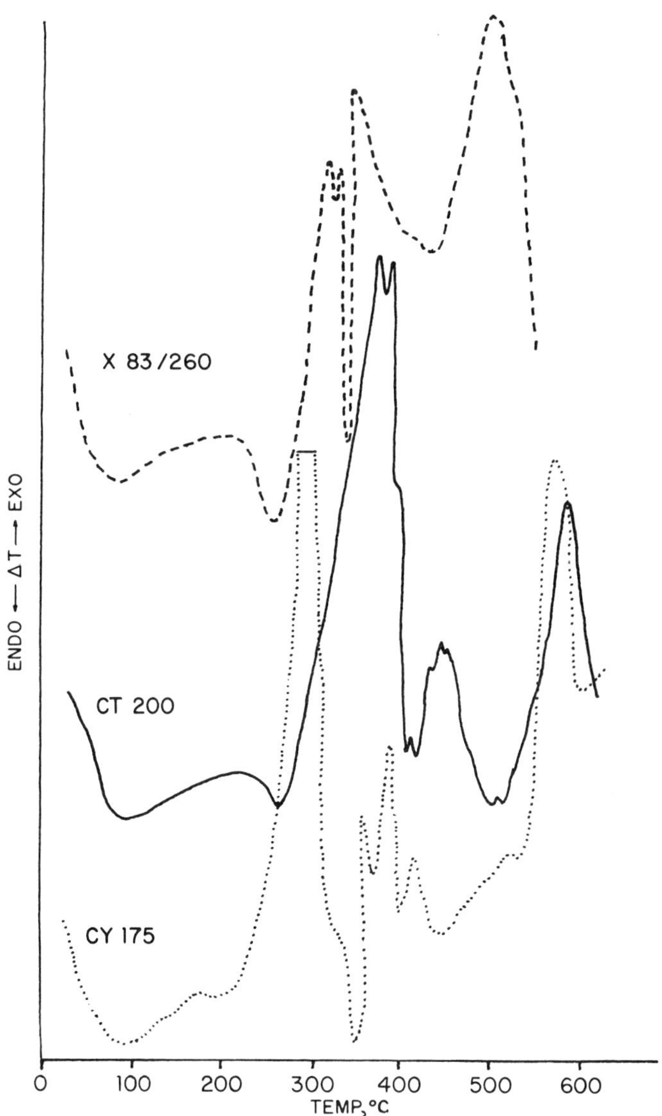

FIG. 1. DTA of uncured epoxy resins; CT200 bisphenol type, CY175 cycloaliphatic type, and X83/260 mixed types.

                                                                    MANLEY

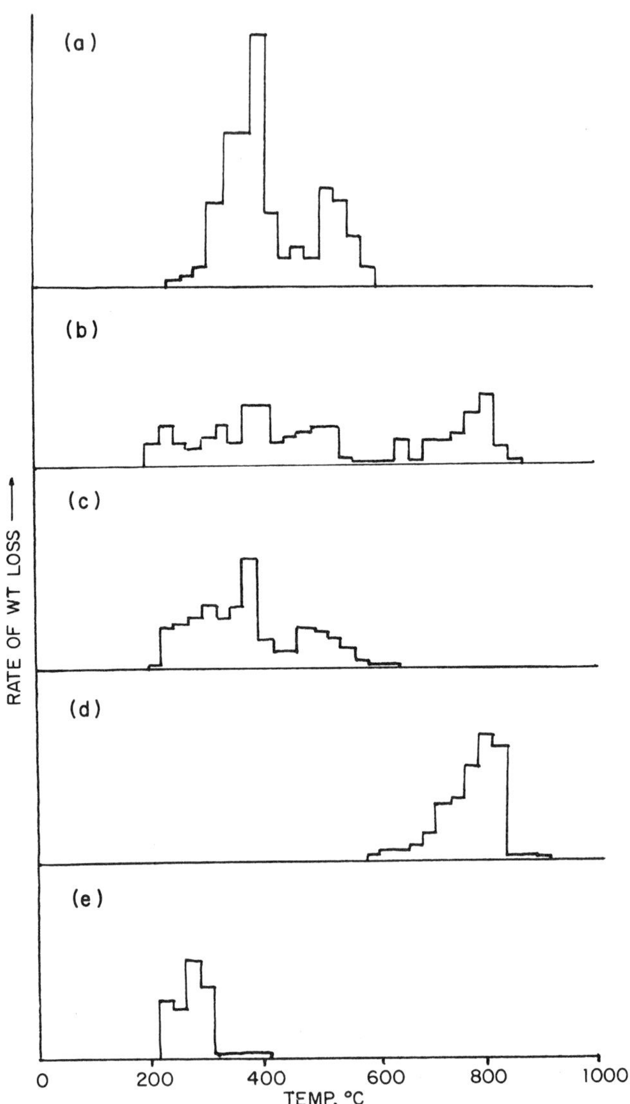

FIG. 2. TG of CT200, (a) Uncured. (b) Cured with HPA and filled with $CaCO_3$. (c) Cured with HPA and filled with $Al(OH)_3$. (d) $CaCO_3$ alone. (e) $Al(OH)_3$ alone.

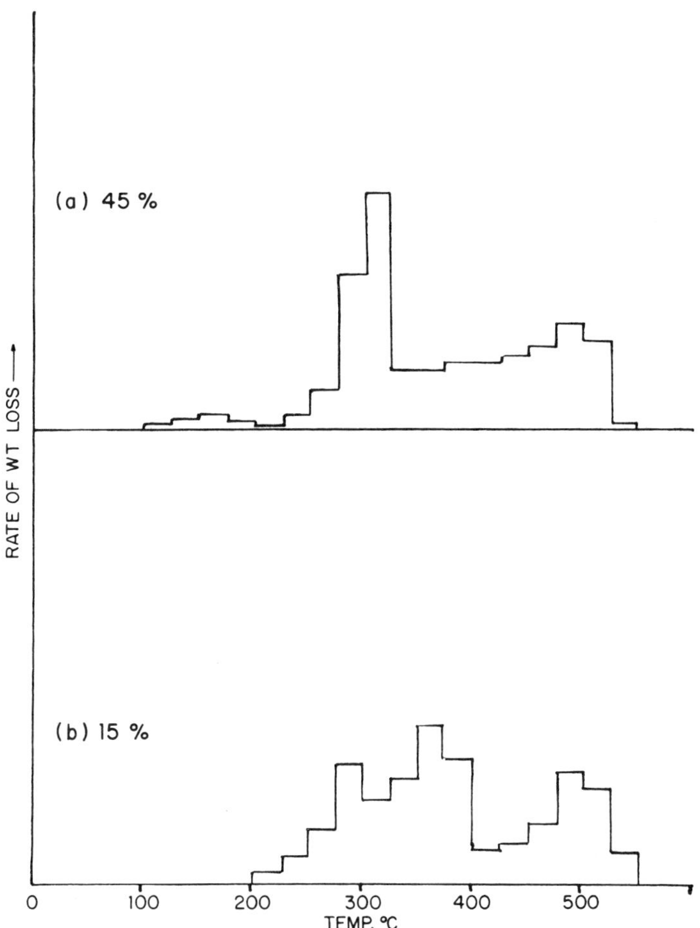

FIG. 3. TG of Epikote 828. (a) With 45% phthalic anhydride. (b) With 15% phthalic anhydride.

molding pressure; there are, of course, no curing endotherms, and it is also noticeable that the characteristic exotherm is sharper (instead of a split peak there is only a shoulder) and at a higher temperature. Curve c shows a curve of cured resin under pressure; here the endotherm following the characteristic exotherm is suppressed, indicating that this trough is caused by volatilization of decomposition products.

When MF resins are molded, excess formaldehyde escapes when

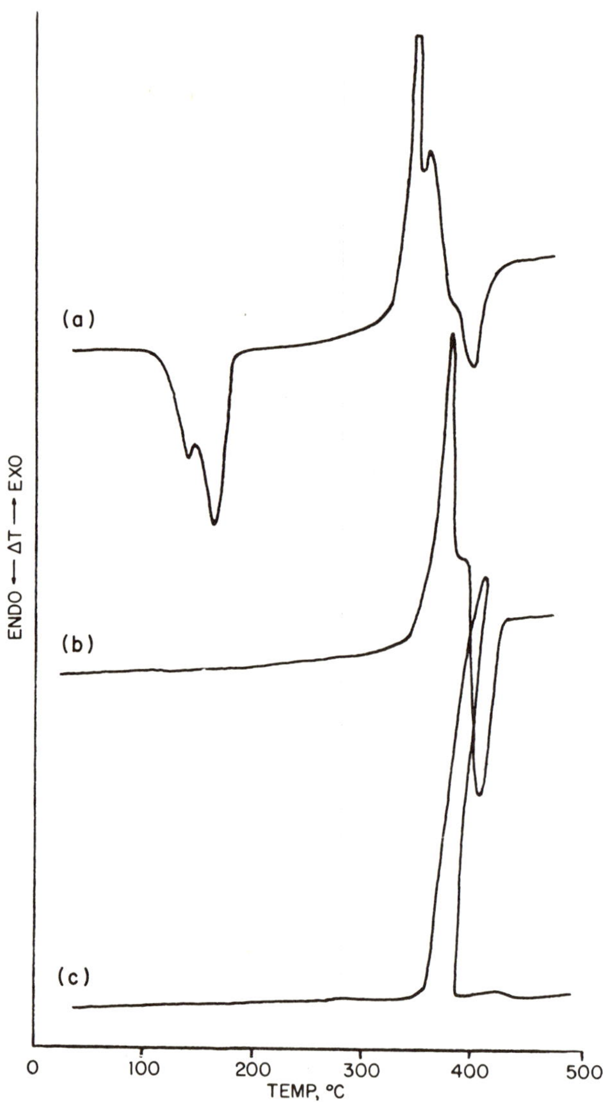

FIG. 4. DTA of melamine formaldehyde resin. (a) Uncured MF
resin in pressure DTA cell. (b) MF resin cured by compression
molding in conventional DTA. (c) Molded resin in pressure DTA cell.

the mold is "breathed," but when cured in a closed DTA cell this HCHO
may react further with the melamine. Commercial conditions therefore favor a resin with methylene bridges as the linking units while the
DTA cured resins have more methylol and ether groups. These latter
are less stable than methylene bridges and are presumed to be
responsible for the lower temperature of the characteristic exotherm.

TG has proved as useful for the characterization of filled amino
resins [2] as it has for the epoxies. Evolved gas analysis (EGD) is a
useful complement to DTA for identifying melamine compounds.
Figure 5a is a DTA trace of hexamethylol melamine resin, and Fig. 5b

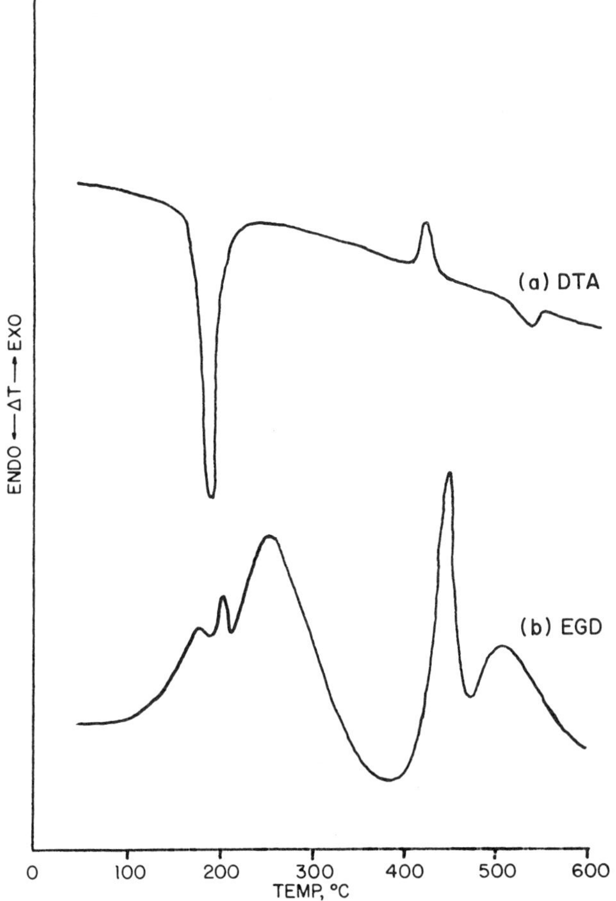

FIG. 5. (a) DTA of hexamethylol melamine resin. (b) EGD of
hexamethylol melamine resin.

is the EGD record showing the loss of water and formaldehyde com-
mencing at $100°C$ and followed by further evolution of gas concomitant
with the curing reaction. The characteristic peak of the melamine
appears just after the DTA peak, and around $460°C$ gas is evolved from
a subsequent reaction. When compared with the conventional DTA
(Fig. 5a), the extra resolving power of EGD is evident. EGD on cured
melamine resins show peaks corresponding to the reactions observed
by May [3] and thus provide a further means of identifying and studying
these materials. When using a gas density detector the $460°$-melamine
peak is seen to have two components, one lighter and one heavier than
argon.

CURING REACTIONS

The curing of thermosetting resins was one of the first applications
of DTA [4], and it has been extensively applied to the study of the
kinetics of the reaction.

One source of error, however, arises from heat transfer from the
specimen to the reference cell. Figure 6 shows the area of exothermic
peak obtained with various sizes of specimen in a Du Pont 900 Thermal
Analyzer operated in the DSC mode. The ratio is only constant for
certain sample sizes, and departures from linearity increase as the
sample size increases. This factor does not seem to have been taken
into account in some reported work on curing of resins with DTA and
DSC. It can easily be avoided in specially designed apparatus [5].

In many cases, however, what is required is not a knowledge of the
full curing process, but an indication of the degree of cure. As the
residual exotherm is small, modified DTA is preferred to traditional
apparatus for this application. One modification is to use a larger
sample; a second is to reduce the rate of loss of heat from the sample
[5] in order to increase the autocatalytic effect; third, one may replace
thermocouples with thermistors [6]. In most instances isothermal
operation is preferred to normal DTA.

When dealing with resins that show a glass transition (T_g), such as
the bisphenol epoxies, another method of determining the degree of cure
is to follow the T_g. This provides a sensitive index of degree of cure
that can be correlated with standard heat distortion temperatures.

Figure 7a shows the values found for T_g of CT200 cured with phthalic
anhydride at $160°$ after gelation at $120°C$ for 24 hr. The curing is substantially complete after 16 hr at $160°$. These results give good correlation with other thermal and classical techniques.

This method is restricted to polymers with rigid chains and is of
much less utility with the normal filled resin of commerce. These both
raise T_g and reduce the sensitivity of the technique.

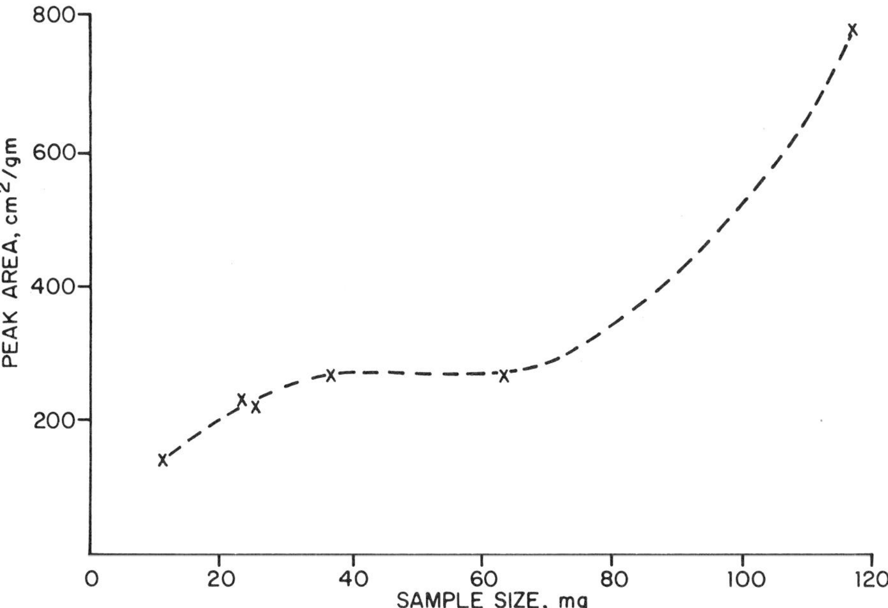

FIG. 6. Area of exothermic peak obtained in curing various sizes of epoxy resin.

As before, however, it is possible to make specialized apparatus with higher sensitivity than that of standard DTA equipment; e.g., T_g may be determined in heat flow apparatus [7].

DEGRADATION REACTIONS

As a practical tool TG is very valuable [8], but it has disadvantages for measurement of absolute kinetic values [9]. Figure 7b shows how T_g measurements may be used to monitor the degradation of epoxy resins.

The degradation of melamine formaldehyde resins has had comparatively little study and the results obtained are conflicting. This is in part because MF resins are used in surface coatings as well as in compression molding. The composition of the resins used are different in each instance, as are the curing conditions and the criteria of degradation. The use of different analytical techniques (DTA, pressure DTA, IR, and TG) also give different results [10].

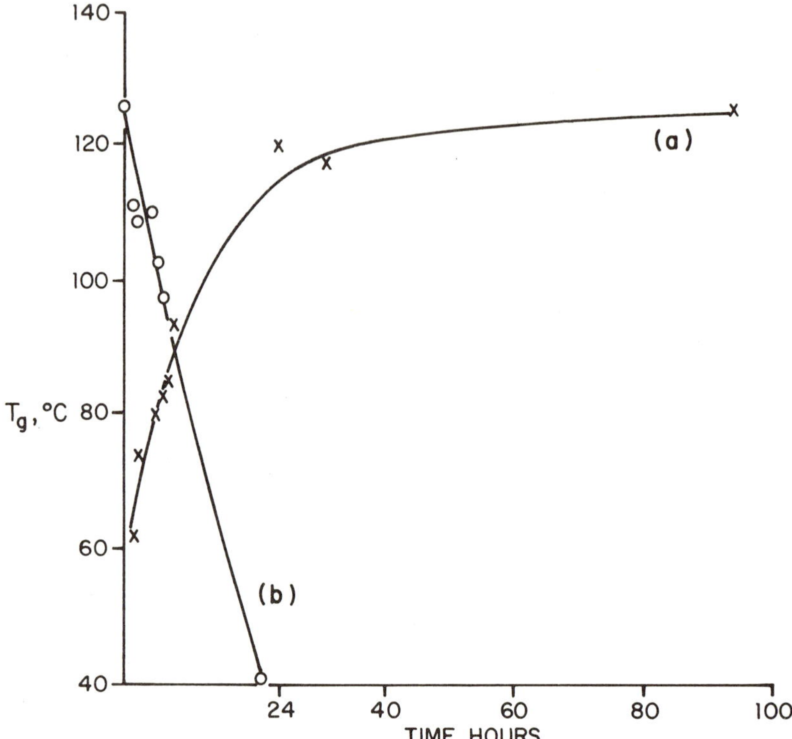

FIG. 7. T_g of CT200 and phthalic anhydride.   (a) After curing at 160° (×).   (b) After heating at 220° (○).

In studies of compression molded compounds using TG [11], a rapid reaction was observed above 250°C which was attributed to breakdown of the triazine ring. It is more probable that this was an oxidation reaction (it was greatly dependent on the particle size of the material investigated). Melamine begins to sublime above 260°C [11] and thus an oxidation of the methylene bridges followed by sublimation of melamine is the most likely cause of the disappearance of the triazine absorption from the IR spectra.

In the absence of oxygen the triazine ring is stable above 500°C [3, 10], and thus another explanation has to be found for this characteristic exotherm.

This exotherm is not found in phenolic resins and therefore by analogy with the reaction in melamine [3] the following mechanism is postulated:

$$RNHCH_2HNR \rightarrow RNHR + [CH_2NH]$$

If the $[CH_2NH]$ were to dimerize to azomethane, this would decompose exothermically at $400°C$ and the possible reactions are legion. The methyl radicals could dimerize to ethane or, if even a trace of oxygen is present, to polyolefins. If air is introduced above $360°C$ a very sharp, large $(+50°C)$ exotherm is produced that is much too rapid for a normal combustion. Reaction is also possible with the NH_3 [3] present to give amines which have been detected by EGD.

All that can be said for certain is that the reactions are too complex to be fully clarified by thermal analysis alone.

EXPERIMENTAL

The MF resins (F/M ratio 2.5 to 1) were cured at $160°C$ for 10 min and ground to B.SS. 200-300 mesh.

A Stanton thermal balance HT-SM and a Mark IIB electrical thermobalance (C. I. Electronics, Salisbury, Wiltshire) were used. Conventional differential thermal analysis (DTA) was performed on the equipment manufactured by A. R. Bolton of Edinburgh based on a design of R. C. Mackenzie and on the Du Pont 900 using the DSC mode. The pressure cell [10] and EGD apparatus [12] have been described elsewhere.

CONCLUSIONS

Thermoanalytical methods are useful in the characterization of thermosetting compounds, but no single technique suffices to all the complex reactions involved. DTA enables one to identify thermosets and to distinguish between filled and unfilled resins, between resins cured in open or closed cells, and between mixtures and copolymers. EGD is a useful complement to DTA for characterization and curing studies. TG is a conventional method of characterizing commercial amino and epoxy resins. Thermal methods alone cannot resolve the complex degradation reactions of thermosets.

Some of the problems in the study of thermosetting resins are found in many thermal investigations, e.g., problems of diffusion and the effects of particle size, but others such as the sublimation of melamine and the possible reactions between volatile condensation products and the main resin are peculiar to this field of study.

REFERENCES

[1] T. R. Manley, in High Voltage Applications of Epoxy Resins (M. Billings, ed.), Univ. Manchester Press, 1967, p. 1.

[2] T. R. Manley, Trans. Plastics Inst., 1967, 525.
[3] H. May, J. Appl. Chem., 9, 340 (1959).
[4] T. R. Manley, SCI Monograph No. 17,
     Society of Chemical Industry, London 1962, p. 175.
[5] T. R. Manley, Med. Biol. Eng., 7, 71 (1969).
[6] D. A. Higgs and T. R. Manley, To Be Published.
[7] T. R. Manley, J. Thermal. Anal., 2, 411 (1970).
[8] D. J. Toop, Trans. IEEE Trans. Elec. Insul., 6(1), 2 (1971).
[9] T. R. Manley and C. G. Martin, Acta Chim. Acad. Sci. Hung.,
     67(4), 443 (1970).
[10] T. R. Manley, IUPAC Symposium on Macromolecules, Preprint
      IV-20, 1972.
[11] W. R. Moore and E. Donnelly, J. Appl. Chem., 13, 537 (1963).
[12] T. R. Manley and D. A. Higgs, J. Appl. Polym. Sci., 16, 1039
      (1972).

Characterization of Aging Skin via Thermal Analysis

W. T. HUMPHRIES and R. H. WILDNAUER

Department of Skin Biology
Johnson & Johnson Research
New Brunswick, New Jersey 08903

ABSTRACT

Samples of full-thickness rat skin from birth to 3 years
were characterized by thermal analysis. The techniques
involve differential scanning calorimetry, thermomechanical
analysis, and thermogravimetry.
In the dry state the skin samples exhibit thermally induced
transverse dimensional changes at 33, 130, 192, and 250°C.
The 130 and 192°C expansions are complimented by
enthalpy changes. As the animal ages, the degree of the
individual expansion decreases as the energy associated
with the transition increases. This observation is consistent with the theory of increased cross-linking with age.
The relationship observed in the dry state is also present
in hydrated samples.
Thermogravimetric results indicate an initial weight loss
up to 100°C, which is due to dehydration, followed by a
major weight loss between 250 and 450°C. The decomposition products in the higher temperature range also varied
with age. Work with recast collagen films and elastin aid
in the interpretation of the thermal transitions observed in
the native rat skin.

INTRODUCTION

The major biophysical properties of mammalian skin are mainly determined by its dermal layer. This connective tissue is a complex fibrous network held together by collagenous and elastin-type fibers. These fibers have been shown to undergo distinct thermally induced structural transitions (e.g., shrinkage phenomena of collagen) and have been the subject of numerous investigations over the past decades [1]. With the aid of x-ray diffraction [2], this shrinkage process in collagen has been assigned to a phase transition involving the conversion of the crystalline helical collagen structure to an amorphous random coil form.

The objective of the present work is twofold: 1) characterization of this biopolymeric system by thermal analytical techniques, and 2) to determine if these thermal characterizations are a function of the age of the animal.

The high degree of organization at both the morphological and molecular level suggests that the dermis should be amenable to thermal techniques commonly used to characterize polymeric systems. A similar approach has been taken to better understand the structure property relationships of other biorelated tissues such as hair and stratum corneum [3-5].

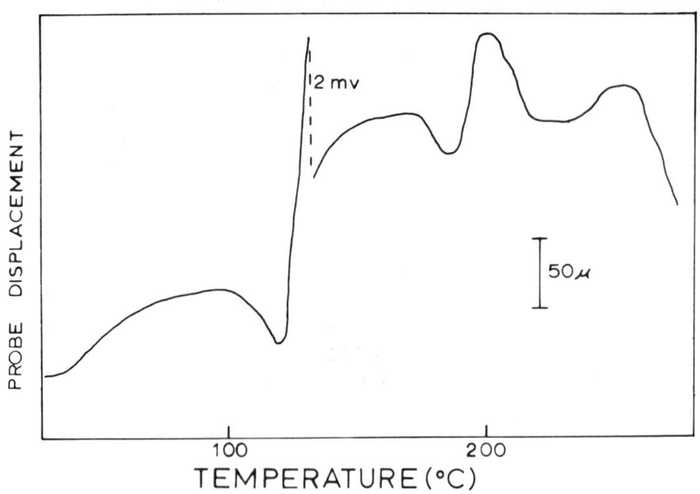

FIG. 1. Thermomechanical analysis of 10-day old rat skin.

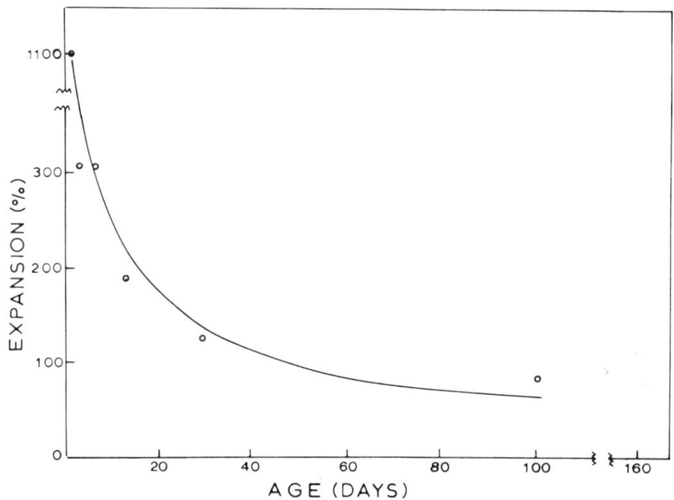

FIG. 2. Transverse expansion at $190°C$ vs age for dry material.

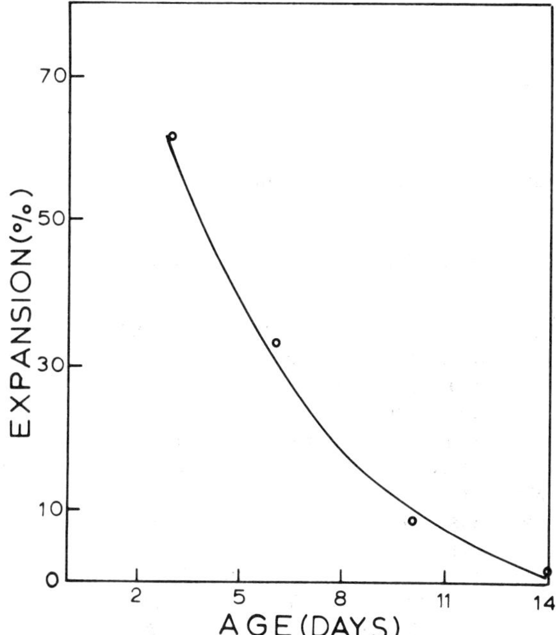

FIG. 3. Degree of $33°C$ transition vs age.

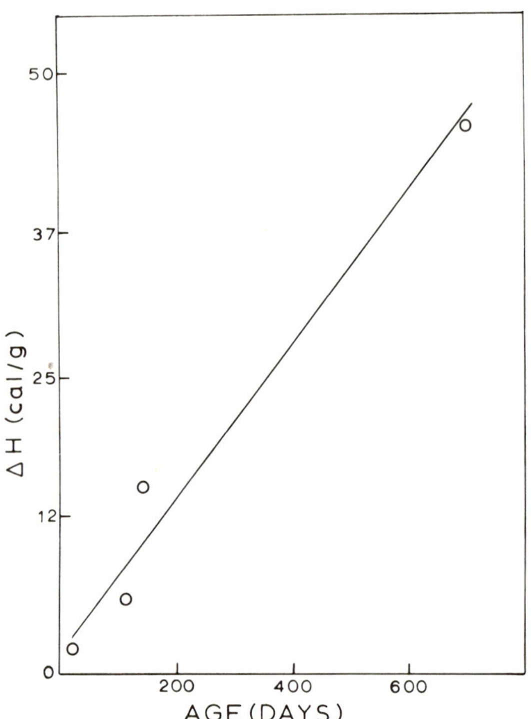

FIG. 4. Relationship between ΔH and age for dry samples.

MATERIALS AND METHODS

The skin samples were obtained from a controlled colony of Charles
River CD strain rats. Animals were sacrificed at various times from
birth up to ages approaching their normal life span (2 to 3 years). The
hair from the skin was removed mechanically without the aid of a
depilatory agent. The samples were stored in a 0.9% by weight NaCl
solution at 35°F or dried to room temperature prior to examination.
Storage in the NaCl solution for periods of up to 5 days did not alter
the results. The recast collagen films were prepared from a ficin
digested Bovine tendon.
The thermal techniques involved the use of a Du Pont 990 thermal
analysis system equipped with a differential scanning calorimeter cell
as well as the Du Pont 951 Thermogravimetric Analyzer. The thermomechanical analysis was performed with a Perkin-Elmer Model TMS-1

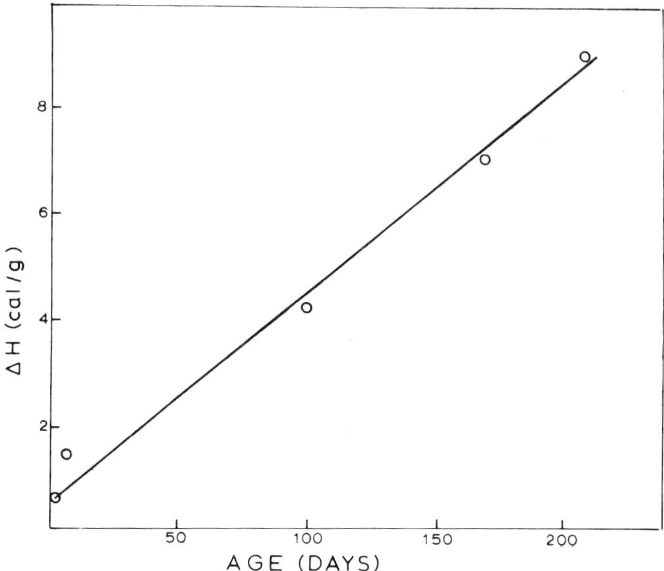

FIG. 5. Agreement between ΔH and age for wet samples.

Thermomechanical Analyzer. The TMA and DSC evaluations were
performed in a helium atmosphere while the TG was in nitrogen.
   In the TMA, the expansion mode was utilized with a zero applied
force. The magnitude of the transverse expansion was calculated
based on the initial sample thickness.

EXPERIMENTAL RESULTS

   In the thermomechanical analysis, the dried skin samples, exhibited four thermally induced transverse dimensional changes: 33,
130, 192, and 250°C. A TMA thermogram for a 10-day old rat is
shown in Fig. 1. In this case the total degree of expansion at 250°C
is 300% of the sample's original thickness. In previous work with
a recast collagen film, only the 190°C transition was observed in the
dry state. If the recast collagen film was exposed to a diluent such
as silicone oil, the 190°C transition was reduced to 137°C. This
lowering of melting temperature of collagen by diluents was observed
by Flory [6]. By altering the diluent concentration, Flory was able to
extrapolate to the melting point of the dry material.

These two collagen transitions (130 and 190°C) are both present in the rat skin. The magnitude of the total expansion of both transitions does decrease considerably with the age of the animal. The relationship between the expansion and the age is shown in Fig. 2.

The transition observed at 33°C was absent in the recast collagen film and may be a result of some other component (e.g., lipids) or a low molecular weight fraction of the collagen which is not present in the recast film. The relationship between age and the degree of this 33°C transition is shown in Fig. 3. This initial transition decreased more rapidly than the 190°C transition and was completely absent by the twentieth day. A pellet prepared from isolated elastin extracted from ligamentum nuchae exhibited only one thermally induced transverse expansion at 250°C. This transition temperature corresponds to the high temperature expansion observed in the rat skin.

Samples of fresh rat dermis examined in 0.9% NaCl solution exhibited only one transverse expansion at 70°C. This temperature corresponds to the normally reported shrink temperature of

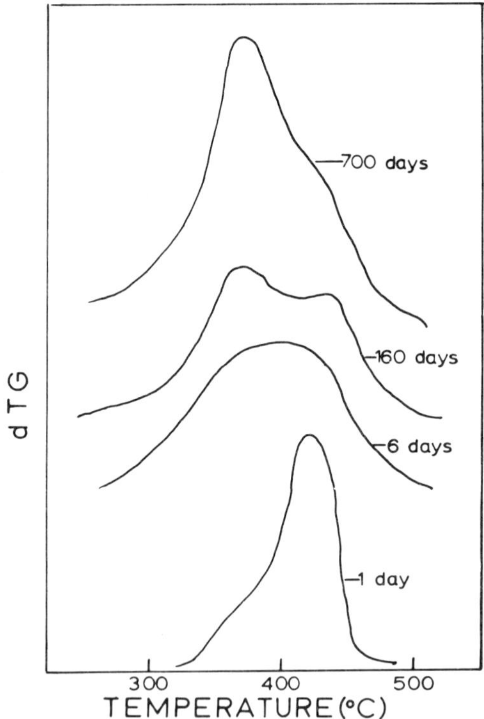

FIG. 6. dTG as a function of temperature for dry samples.

collagen. The magnitude of the expansion decreased with age but not as significantly as the dried material. The rat dermal samples contained equilivate amounts of water as determined by drying the samples to room conditions followed by thermogravimetric analysis.

In the differential scanning calorimeter, the dried skin samples exhibited a broad endotherm with a minimum at 70°C, which is probably the loss of water, followed by sharper endotherms at 130 and 180°C. The high-temperature endotherms agree with the dimensional changes observed in the thermomechanical analysis. The ΔH associated with the 130 and 180°C endotherms increased dramatically with age. The agreement between ΔH and age of the animal is shown in Fig. 4. The relationship between ΔH and age is considerably more linear than the TMA data. As in the TMA, the recast collagen film only exhibited the 190°C transition.

The animal samples stored in 0.9% NaCl solution were run in hermetically sealed pans to prevent the loss of solvent. The DSC thermogram showed two distinct endotherms with minimums at 62 and 75°C. As with the dry sample, the total ΔH observed increased with age. The ΔH as a function of age is shown in Fig. 5.

The thermogravimetric analysis on the dried animal skin exhibited a gradual weight loss from 25 to 100°C representing the loss of water (~10%) and a more substantial loss between 250 and 450°C. A plot of the differential of the weight loss vs temperature between 250 and 450°C did expose some variation with age. The derivative thermogravimetric (dTG) curves are shown in Fig. 6. The younger samples contained only a higher temperature component. As the animal aged, a second component at a lower decomposition temperature appeared. A further analysis of the evolved gas is needed to distinguish between these components.

CONCLUSION

Modern thermal analytical techniques are useful in the study of the structure property relationship of these biopolymeric systems. The techniques can accommodate materials in the dry or wet state as well as only requiring very small samples, which is usually a major problem in examining these biopolymers.

This work is highly descriptive but does illustrate that thermal analysis is a tool which reveals changes in macromolecular structure with age. The reduction in the degree of expansion in the TMA and the increase in the ΔH associated with the transitions are consistent with the theory of increased cross-linking with age. Since the dermal system is a complex mixture, further biochemical work in connection with these thermal techniques is necessary to establish the macromolecular events occurring during the observed transitions and what type of cross-links, if any, are changing with age.

REFERENCES

[1] A. Veis, Treatise on Collagen (G. N. Ramachandran, ed.),
     Academic, London, 1967.
[2] D. Puett, A. Ciferni, and L. V. Rajagh, Biopolymers, 3, 439
     (1965).
[3] W. T. Humphries and R. H. Wildnauer, J. Invest. Dermatol.,
     57, 32-37 (July 1971).
[4] W. T. Humphries and R. H. Wildnauer, Ibid., 58, 9-13
     (January 1972).
[5] W. T. Humphries, D. L. Miller, and R. H. Wildnauer, J. Soc.
     Cosmet. Chem., 23, 359-370 (May 1970).
[6] P. J. Flory and R. R. Garrett, J. Amer. Chem. Soc., 80, 4836
     (1958).

Advances in Thermogravimetric Analyses
of Elastomer Systems

JOHN J. MAURER

Enjay Polymer Laboratories
Linden, New Jersey 07036

ABSTRACT

Practical rubber formulations are complex mixtures of
polymer(s), fillers, plasticizers, curatives, and processing aids. Classical methods exist for determining many
of the components of these systems but are too time
consuming for routine use in quality control or problem
solving activities. Thus a need has existed for more
rapid procedures. Previous studies of gum, compounded,
and cured elastomers showed that thermogravimetry (TG)
can provide information regarding the polymer, oil, carbon
black, mineral filler, and ash content of vulcanizates. The
present study extends this approach to multicomponent
polymer and carbon black systems to determine whether
components of similar thermal characteristics can be ef-
fectively analyzed. Vulcanizates containing binary (natural
rubber (NR)/ethylene propylene terpolymer (EPDM) or
ternary (NR/EPDM/styrene-butadiene rubber) blends were
successfully analyzed by the standard, dynamic TG method.
Isothermal TG was shown to have significant potential for

analyzing blends of FEF and SRF carbon blacks (CB) in a standard formulation. Two additional approaches, derivative TG (DTG) and normal probability analysis are examined for potential applicability in these analyses. The former has apparent value for polymer blend analysis; the latter, based on preliminary data, may have value in the analysis of some CB blends.

RESULTS AND DISCUSSION

Analysis of Polymer Blends in Vulcanizates

Evaluation of polymer decomposition characteristics has potential for quantitative analyses of polymer blends. The degree to which this can be achieved in a given elastomer system will depend on a variety of factors including the overall composition of the rubber formulation as well as the relative inherent characteristics of the particular polymers. The general problem to be considered is the possible overlap of the thermal decomposition or volatilization characteristics of the components of the formulation. When this occurs due to nonpolymeric components, it can often be compensated for by extraction of the sample prior to thermogravimetry by a correction procedure for a "standard" system, or by manipulation of the TG conditions to selectively remove volatile, nonpolymeric components prior to decomposition of the polymers [1]. The following discussion is concerned, however, with another type of overlap problem, i.e., that due to the polymer components in a blend.

Two-Component Blends

Assuming no interaction between the polymeric components, quantitative analysis via TG should be feasible for many elastomer blends if the components differ to a high degree in thermal stability. The present case deals with a more difficult system wherein 1) one of the components (natural rubber, NR) has a more complex decomposition pattern than the other (ethylene propylene terpolymer, EPDM), and 2) the systems overlap to a moderate degree. Figure 1 compares the individual systems and a 50/50 blend of the two polymers. Initial inspection of the shape of the TG curve for the latter system does not suggest a major difference vs that of the 100 NR system. However, it will be shown that careful examination of both the position and the detailed shape of the TG curve can provide an effective basis for quantitative analysis of polymer composition in these blend systems.

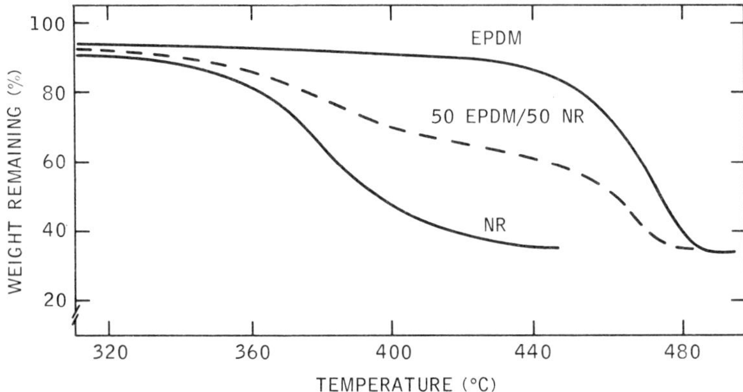

FIG. 1. TG analysis of EPDM/natural rubber systems.

The first method by which such analysis can be accomplished takes advantage of the fact that most of the NR decomposition occurs at a lower temperature than that of the EPDM system. Thus it is possible to locate a temperature above which most of the polymer weight loss is due to EPDM. A graph of the data in Table 1 reveals a highly linear relationship. This system also presents an opportunity for evaluating the derivative of the TG curve as a means to amplify small, but significant differences in the detailed shape of primary TG curves such as those shown in Fig. 1. Notice, for example, that the derivative thermogravimetry (DTG) curves (Fig. 2) make it much easier to distinguish between these systems. Table 1 compares the peak heights of such

TABLE 1. TG Analysis of Vulcanized Natural Rubber/EPDM Blends

Composition wt% EPDM	Wt% remaining (410°C)	DTG Peak (in.) (NR)
0	8.5	2.95
20	26.3	2.20
50	48.8	1.30
80	63.3	0.78
100	84.7	-

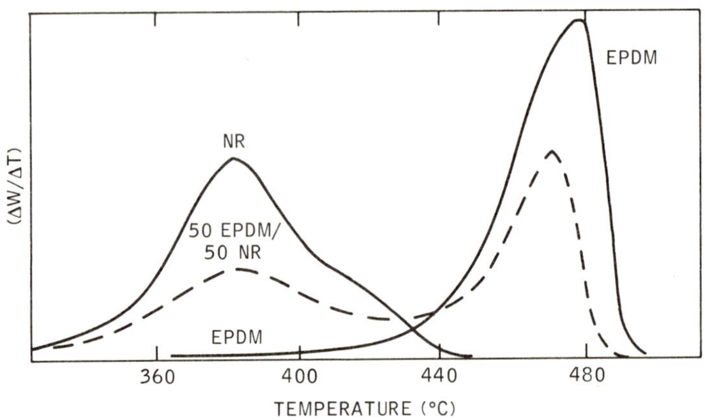

FIG. 2. Derivative TG of EPDM/NR systems.

curves to blend compositions; a graph of these data indicates that this
approach, properly calibrated, should be feasible. An interesting
aspect of this type of system is that there are at least two measures of
polymer composition to provide a cross-check of polymer content in
the blend.

Three-Component Blends

The success of the methods outlined above for the two component
systems prompted examination of their applicability to an even more
complex system, in this case, ternary blends of NR, styrene-butadiene
rubber (SBR), and EPDM. The particular blends under examination had
a fixed EPDM content of 20 phr; NR and SBR contents ranged from zero
to 80 phr. Figures 3 and 4, which present representative TG and DTG
data for these systems, reveal basic characteristics generally similar
to those of the NR/EPDM system. Plots of the weight loss and peak
height data of Table 2 indicate that quantitative analysis via TG techniques is feasible even for these complex systems.

Due particulary to the complexity of the systems described throughout this paper, it is essential that the basic TG experiment be conducted
in a standardized, highly controlled manner [1, 2]. It has previously
been established that meaningful analyses of vulcanizate composition
can be conducted if reproducible operating conditions are established,
monitored, and maintained. Although the methods outlined in this
presentation have been developed for specific vulcanizates, it is
reasonable to anticipate that other compositions may be analyzed in
similar fashion. Because of the wide variety of practical formulations
which may be encountered, however, each system must be treated as a

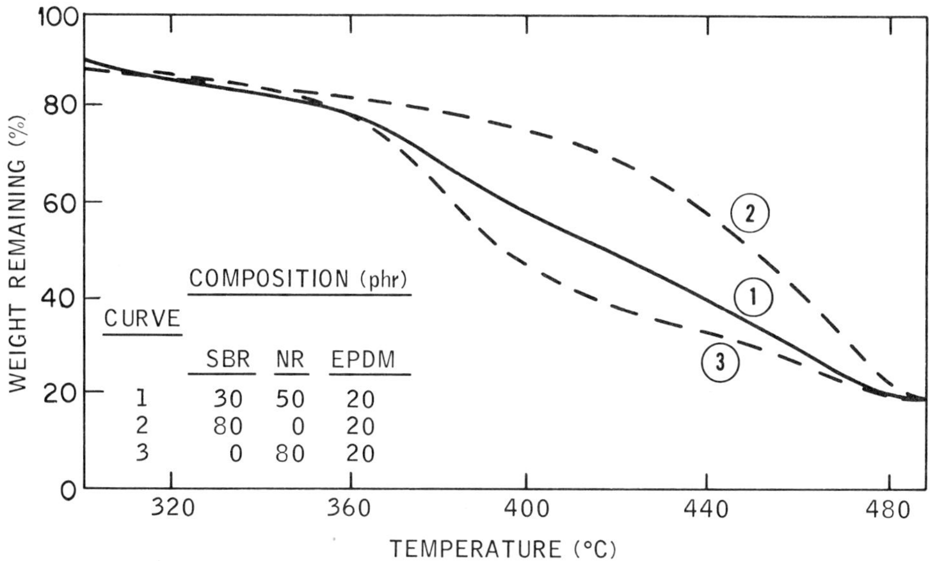

FIG. 3. TG analysis of SBR/NR/EPDM systems.

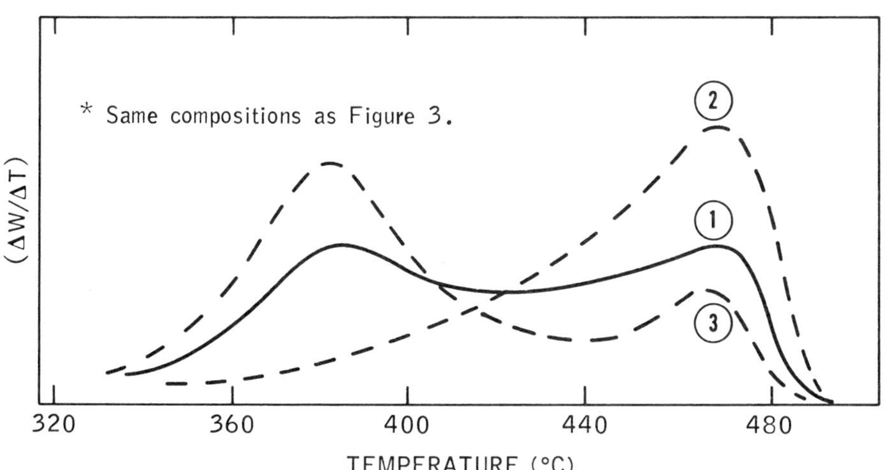

FIG. 4. Derivative TG of SBR/NR/EPDM systems.

TABLE 2. TG Analysis of Vulcanized EPDM/NR/SBR Blends

Composition (phr)			Weight % remaining (400°C)	DTG peak (in.)	
EPDM	NR	SBR		NR	SBR
20	80	0	40.5	2.81	1.35
20	50	30	58.0	1.85	1.85
20	30	50	66.5	1.28	2.35
20	0	80	80.0	0.60	3.25

new case until the feasibility of these methods have been established
for the particular formulation of interest.

Analysis of Carbon Black Blends in Vulcanizates

Practical rubber compounds frequently contain blends of carbon
blacks (CB) of various particle size, surface area, and/or "structure"
chosen to yield the proper balance of compound cost and product
properties. It has previously been shown that single carbon blacks
of different surface area could be statistically differentiated in a
standard vulcanizate via their oxidation characteristics during controlled TG experiments. The potential for analyzing CB blends whose
components had large differences in surface area was also indicated.
The present treatment will examine the applicability of such techniques to blends of similar surface area. Before proceeding into a
detailed analysis of one of these systems, mention should be made of
the fact that previous work indicated that both physical and chemical
factors associated with the complete formulation can influence the
decomposition of carbon blacks during this type of analysis [2]. The
present case will involve only factors associated with the carbon
blacks since all of the compounds are in a fixed formulation except
for CB blend ratio.

Analysis of FEF/SRF Blends

A series of blends was analyzed in the vulcanized, unextracted state
as indicated in the Experimental Procedures. Figure 5 presents the
average data for four determinations of each system. It is apparent that
major variations in the CB blend composition of this formulation should
be readily detected by the TG procedure. Small variations, especially
in blends rich in one component, appear directionally more difficult to

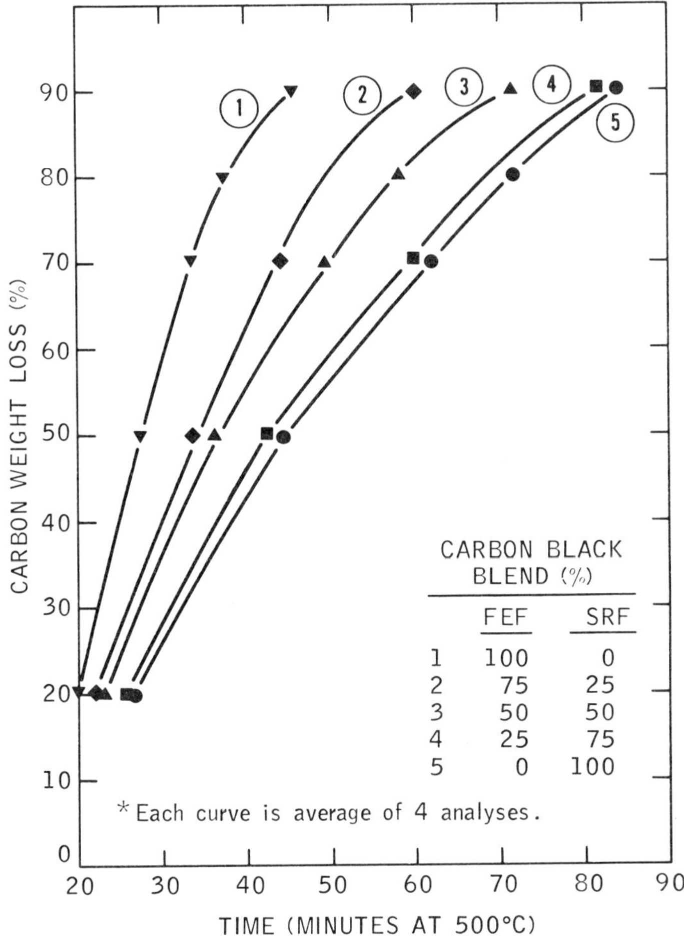

FIG. 5. Analysis of SRF/FEF blends in vulcanizate residues.

detect. Replicate determinations will probably be required for successful analyses of such systems. The times to reach various percentages
of total carbon black weight loss have been shown for each system to
illustrate the fact than an improved analysis will probably result from
the use of times to reach a high degree of weight loss. Presumably this
is due to the fact that the residue in this region is primarily some of the
SRF black. Thus, despite their similarity in surface area, these blacks
are, in effect, "separated" during the TG experiment.

Graphical Treatment of CB Oxidation Data

The oxidation characteristics (weight loss vs time) of FEF in
this vulcanizate residue suggest an additional, perhaps novel approach
to CB blend analysis. In particular it was noted (Fig. 6) that
the weight loss curve for this system displayed a high degree
of symmetry. Because of this, it was expected that this curve
could be transformed to linear form via normal probability
analysis. A successful test of this assumption is shown for the
FEF system in Fig. 7. In making this type of plot it is necessary
to have values for the initial and final weights of the material
which was subjected to the oxidation treatment. The initial values
of these quantities were read directly from the TG chart. Minor
adjustment of the value of the final weight was then made to
improve the linearity of the probability plot. The linearity of
this plot suggests that graphical or computer analysis of carbon
black bend composition may be possible for carbon blacks whose
individual oxidation characteristics resemble that of FEF (Fig. 7).
This follows from consideration of the characteristics of normal
distributions and bimodal blends of such systems. In brief, one
would expect that for two CB's which display the FEF type of
oxidation characteristics, the individual components may appear as
two different lines on a normal probability plot; whereas a blend
of these blacks would fall between, and asymptote, these lines.

The SRF data in Fig. 7 show an interesting result, i.e., this 100%
SRF system yields a normal probability plot which resembles that
anticipated for a bimodal blend of normal distribution. Several
possibilities follow from this observation. First, SRF/FEF blends
cannot be simply analyzed by the normal probability approach, but
modification of the method may be useful for a well-calibrated system.
Second, SRF oxidation characteristics follow a much different form
than those of FEF. Third, SRF behaves like a blend of materials,
and this graphical approach may provide a means of estimating the
relative proportion of these components in a given sample. The
latter possibility warrants further consideration since CB oxidation via a dynamic TG method also suggested that the SRF oxidation
pattern was inherently more complex than that of several other
carbon black types [3].

Additional examination of this approach will consist of 1) attempting
to identify another carbon black which resembles the behavior of
FEF, 2) examining blends of these blacks to determine whether the
normal probability analysis route offers any advantages for CB blend
analysis, and 3) some further consideration of the significance of the
SRF behavior depicted in Fig. 7.

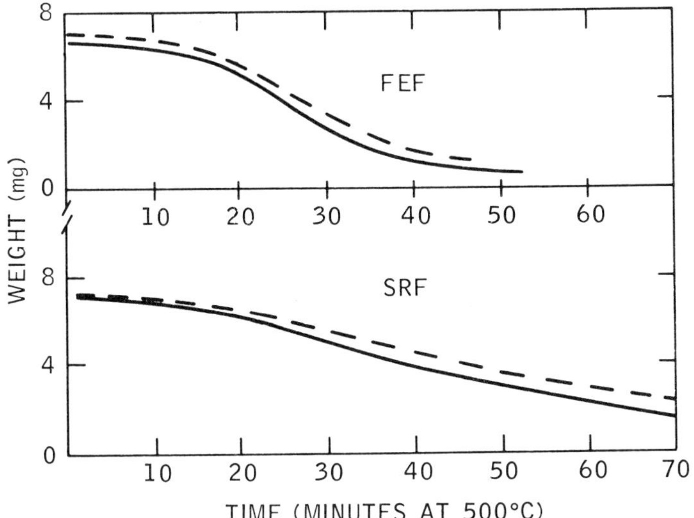

FIG. 6. Oxidation of CB residues from vulcanizates (reproducibility).

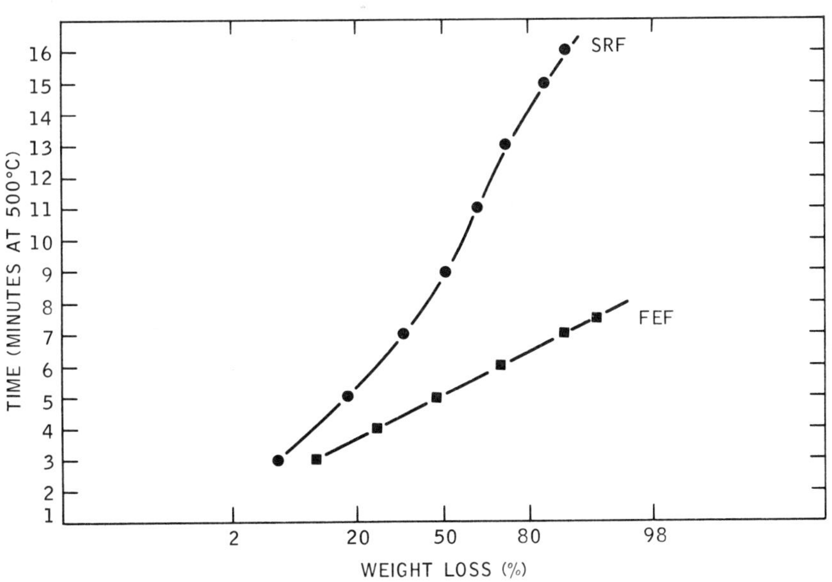

FIG. 7. Decomposition of carbon blacks from vulcanizates: probability plot.

EXPERIMENTAL PROCEDURES

Carbon Black Blends

Formulation. Butyl rubber, 200; carbon black, 100 [FEF (Phil-black "A") and/or SRF (Pelletex NS)]; Stearic acid, 4.0; ZnO, 10.0.

Acceleration. 50 g of above formulation plus 0.40 g sulfur, 0.48 g TMTDS, 0.32 g MBT. Cure: 40 min at 320°F.

TGA conditions. 950 TGA plus Du Pont 900 DTA system. Unit initially purged with N_2 for 15 min. Sample size: 20.2 mg. Remove polymers, etc. by heating to ~550°C in nitrogen; set unit at 500°C (isothermal); switch to air atmosphere; and record weight loss vs time.

Polymer Blends

TGA conditions: Du Pont 950 TGA plus 990 DTA system. Sample size: 6.5 ± 0.3 mg. Nitrogen atmosphere. Heat 10°C/min, dy [(mg/min)/in.] = 0.2. TGA: 1.0 mg/in. Time constant: 1.0. Suppression: 0

EPDM/NR/SBR blends. Composition (phr): Oil (Flexon 580), 10; HAF (Philblack 0), 40; stearic acid, 1.0; ZnO, 5.0; MBT (Captax), 1.5; DPG, 0.75; sulfur, 2. Cure: 30 min at 307°F.

EPDM/NR blends. Composition: Polymer (100); MT, 60; LSF, 20, ZnO, 5; HSt, 2; Santowhite 127, 3; Flexon 840, 10; Altax, 1.5; Tuads, 0.2; sulfur, 1.2. Cure: 15 to 30 min at 307°F.

REFERENCES

[1]   J. J. Maurer, Rubber Chem. Technol., 42, 110 (1969).
[2]   J. J. Maurer, Nat. Bur. Stand. Spec. Publ., 338, 165–185 (October 1970).
[3]   J. J. Maurer, Rubber Age, 102, 47 (1970).

Thermogravimetric Analysis of
Vinyl Chloride/Acrylonitrile Copolymers

B. L. JOESTEN and N. W. JOHNSTON

Chemicals and Plastics Division
Union Carbide Corporation
Bound Brook, New Jersey 08805

ABSTRACT

The dehydrohalogenation of several alternating and random
vinyl chloride/acrylonitrile copolymers was characterized
by thermogravimetry. The polymers were made in solution,
and the conversions were kept below 5% to insure uniform
sequence distributions. Hydrogen chloride was generated
within a relatively narrow temperature range somewhere
between 200 and 300°C depending on the sequence distribution,
relative viscosity, and composition. The weight-loss during
the dehydrohalogenation could be attributed completely to the
hydrogen chloride available in the copolymer. Alternating copolymers were significantly more stable than random copolymers at the same relative viscosity. For a given sequence
distribution or composition, stability decreased with decreasing
relative viscosity. The stability decreased as acrylonitrile
content was increased from 23 to 57%.
The thermogravimetric analyzer was interfaced with a digital
computer. The digitized data were smoothed and differentiated
by convoluted integers. The differentiated data provided rates
for a qualitative discussion of dehydrohalogenation kinetics.

INTRODUCTION

Vinyl chloride/acrylonitrile copolymers are interesting candidates
for studying dehydrohalogenation. They are easily polymerized. In
contrast to the vinyl acetate unit in vinyl chloride/vinyl acetate copolymers, the acrylonitrile unit in vinyl chloride/acrylonitrile copolymers is not expected to generate volatile fragments at the temperatures where dehydrohalogenation occurs in polyvinyl chloride. If so,
vinyl chloride/acrylonitrile copolymers would provide an opportunity to
study dehydrohalogenation in different molecular environments (e.g., by
changing sequence distribution) without interference from other volatile
fragments. Alternating copolymers are known to have different mechanical properties than corresponding random copolymers [1, 2], so it
is of interest to know how sequence distribution affects thermal stability.

EXPERIMENTAL

An alternating copolymer of vinyl chloride/acrylonitrile (VCl/AN)
was prepared using a 1:1 molar ratio of ethyl aluminum dichloride
(EADC) to AN and a 15:1 molar ratio of VCl to AN. This mixture was
polymerized in toluene at $-78°C$ under N_2 atmosphere and yielded a
polymer with an RV of 0.38 in acetone. N, C, and Cl analysis indicated
the polymer (A-1) to be 47 mole % or 43 wt% AN. A higher molecular
weight alternating copolymer of VCl/AN was prepared using a 1.2:1
molar ratio of EADC to AN and a 5:1 molar ratio of VCl to AN. This
mixture was polymerized in toluene at $-78°C$ under N_2 atmosphere
and yielded a polymer with an RV of 1.10 in acetone.

Several random VCl/AN copolymers were prepared by azobisisobutyronitrile (AIBN) initiation in 2-butanone at $65°C$. Reaction times
and AIBN concentration were varied to obtain various molecular
weights. Polymerization data on the alternating and random copolymers are listed in Table 1. All conversions were kept below 5%
in order to insure that all polymers had uniform sequence distributions.

THERMAL STABILITY MEASUREMENTS

Thermal stability was determined by thermogravimetric analysis
(TGA) or thermogravimetry using a Perkin-Elmer TGS-1 Thermobalance.
The TGS-1 incorporates a Cahn electrobalance to measure the weight
of a sample continuously as it is heated according to some selected
thermal program. The sample is heated by a microfurnace that is
calibrated with a set of magnetic metals which lose their magnetism at
well-defined temperatures [3]. The magnetic metals are heated under
the same conditions as the sample to obtain a set of corrections to the

TABLE 1. VCl/AN Copolymers Polymerization

| No. | Polymerization technique | VCl (wt%) | | RV |
		Feed	Copolymer	
A-1	EADC	97	57	0.38
A-2	EADC	86	55	1.10
R-1	AIBN	78	43	0.71
R-2	AIBN	90	57	0.38
R-3	AIBN	90	60	0.24
R-4	AIBN	96	74	0.30
R-6	AIBN	98	77	1.20

temperature indicated on the programmer dial. The set of corrections is fitted by least squares polynomials to a quadratic expression to obtain the corrections at any other temperature.

Approximately 2 mg of polymer were cut from pellets which were compacted at room temperature. After purging the sample environment with nitrogen for 20 to 30 min, the sample was heated at $10°C/min$ in nitrogen.

The TGS-1 was interfaced to an 1130 IBM computer with a Perkin-Elmer ADS-VI Analytical Data System. The ADS-VI provides a digital record of the TGA data via a shaft encoder on a Leeds and Northrup Model "W" Potentiometric Recorder. The digital record provides the computer input which can be analyzed and replotted versus corrected temperature.

RESULTS AND DISCUSSION

Figure 1 shows a typical weight-loss curve of a vinyl chloride/acrylonitrile copolymer when heated in nitrogen at $10°C/min$. The first step of the weight-loss curve represents dehydrohalogenation. The residue at the first plateau does not contain any chlorine. By assuming that the first step of the weight loss represents the quantitative generation of HCl and nothing else, the VCl content of the copolymer can be calculated from

$$\% \text{ VCl} = \% \text{ HCl} \times (62.5/36.5) \tag{1}$$

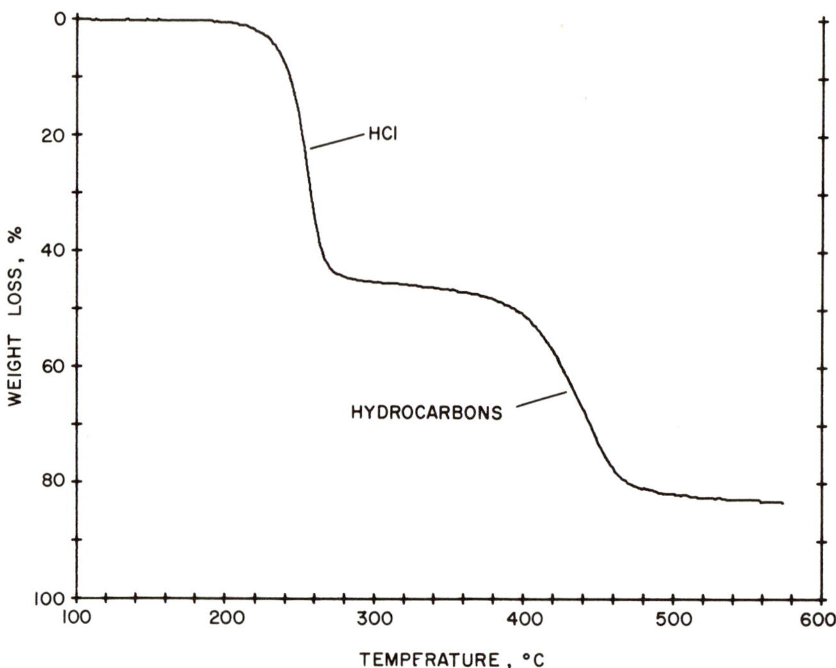

FIG. 1. Weight-loss of typical vinyl chloride/acrylonitrile copolymer heated at 10°C/min in nitrogen.

TABLE 2. VCl Content of VCl/AN Copolymers by TGA

Sample	TGA	VCl (wt%)	
		Cl	C
A-1	56.3	57.1	56.6
A-2	50.5	53.9	55.9
R-1	39.0	41.1	42.8
R-2	54.6	56.0	57.3
R-3	56.2	55.9	59.6
R-4	76.4	76.1	73.7
R-6	78.8	79.3	77.1

Table 2 compares the results of determining the vinyl chloride content from Eq. (1) with the results of chlorine or carbon analysis. With the exception of A-2, the agreement between the TGA results and chlorine analysis is at least as good as the agreement between chlorine analysis and carbon analysis. Therefore, the weight loss in the first step can be completely accounted for by the HCl available in the copolymer. This behavior differs from that of polyvinyl chloride which generates benzene during dehydrohalogenation [4].

The weight-loss data of a random copolymer and an alternating copolymer of the same relative viscosity are compared in Fig. 2. These two copolymers also have approximately the same composition, which is nearly 1 to 1 on a molar basis. The alternating copolymer is considerably more stable than the random copolymer, since it loses weight rapidly about 40° higher than the random copolymer. The improved thermal stability of alternating over random VCl/AN copolymers was briefly noted by Furukawa et al. [5] but no additional data were presented.

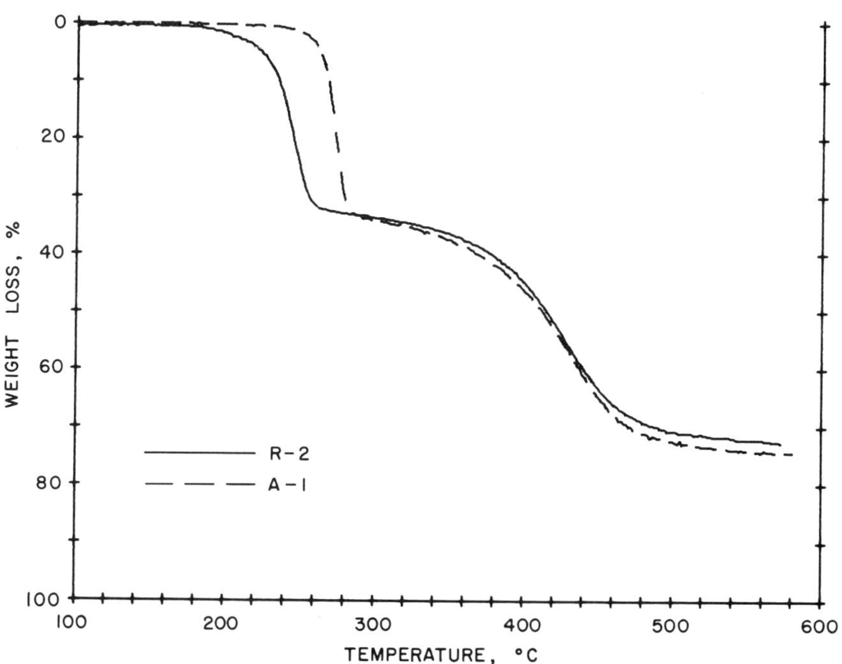

FIG. 2. Effect of sequence distribution on weight-loss of vinyl chloride acrylonitrile copolymers.

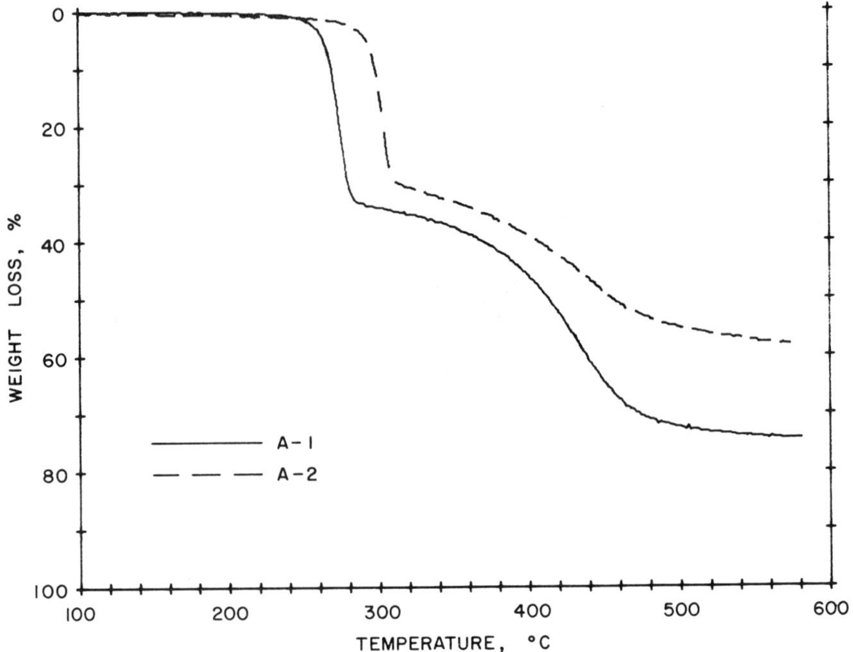

FIG. 3. Effect of relative viscosity on weight-loss of alternating vinyl chloride acrylonitrile copolymers.

The random and alternating copolymers have different sequence distributions. The alternating copolymer should not have any VCl-VCl or AN-AN dyads and, therefore, cannot form allyl chlorides or cyclic nitriles. Copolymer R-2 is a random copolymer and has a predicted sequence distribution [6] of 22% VCl-VCl, 10% AN-AN, and 68% VCl-AN. Films of A-1 and R-2 pressed at the same temperature show A-1 to be colorless and clear while R-2 is yellow-brown.

Figure 3 illustrates the effect of relative viscosity on the stability of the alternating copolymers. The evolution of HCl was delayed about 30° by increasing the relative viscosity from 0.38 to 1.1. The effect of relative viscosity on stability was also observed for random copolymers. Therefore, molecular weight and/or molecular weight distribution also affect the dehydrohalogenation, which suggests that initiation might occur at end groups. A similar effect has been reported for polyvinyl chloride [7].

Based on the chlorine and carbon analysis, A-1 and A-2 have more nearly the same composition than what is suggested by the weight loss in

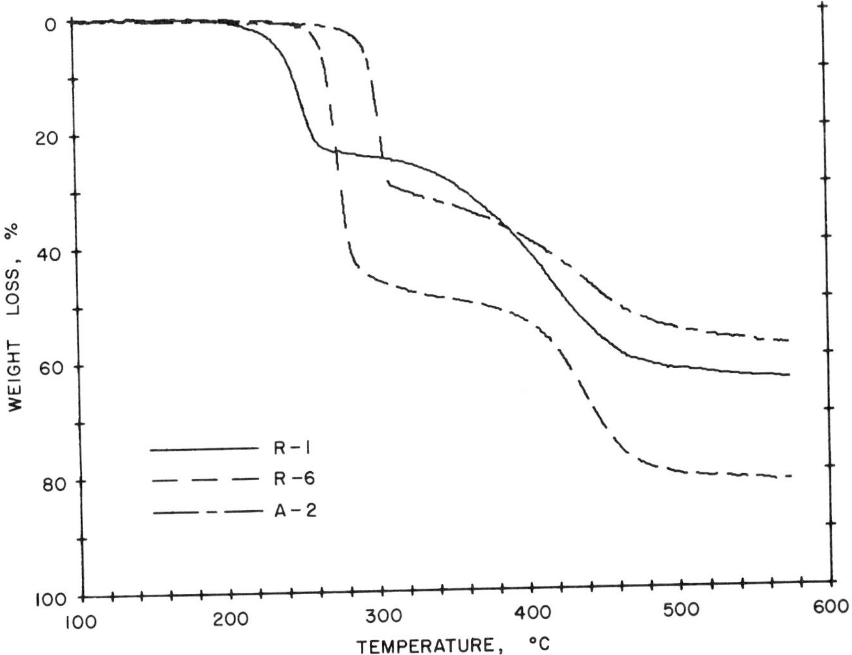

FIG. 4. Effect of composition on weight-loss of vinyl chloride/
acrylonitrile copolymers.

the first step. Recall that A-2 was the exception to the otherwise good
agreement among the three methods for determining vinyl chloride
content. There are evidently some VCl units in A-2 from which HCl is
not eliminated.

Figure 4 illustrates the effect of composition on dehydrohalogenation.
The thermal stability of the random copolymer having 57% AN (R-1) was
less than that of the copolymer having 23% AN (R-6). That is, between
23 and 57% AN, the thermal stability of random copolymers decreases
as the amount of AN increases. However, the deleterious effect of the
AN units can be circumvented if they alternate with VCl units (A-2).

The weight-loss of A-2 and R-6 is compared with that of a polyvinyl
chloride suspension resin (0.85 RV) in Fig. 5. Based on this comparison
the alternating copolymer is probably at least as stable as PVC, and
perhaps more stable, depending on how much the stability of PVC is
affected by relative viscosity.

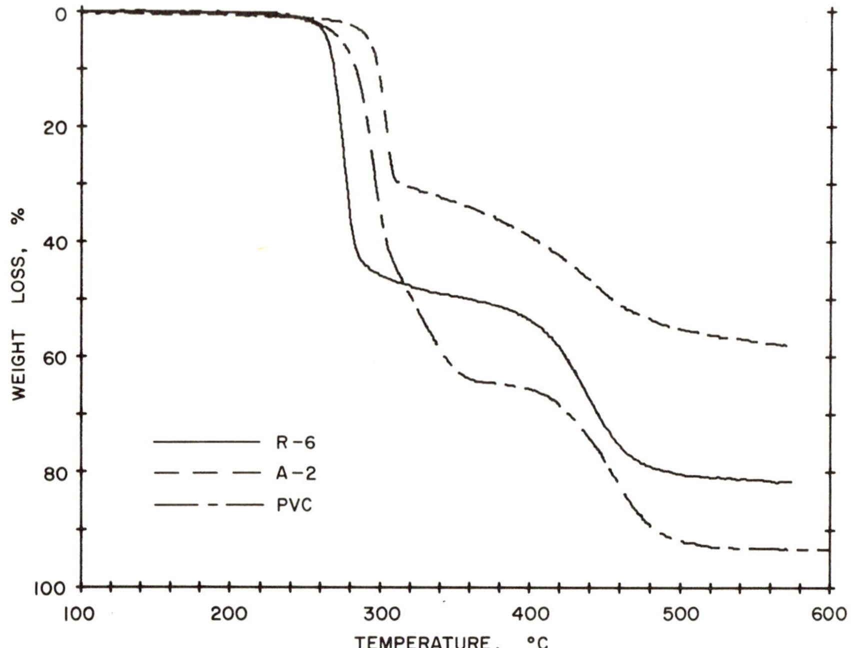

FIG. 5. Weight-loss of vinyl chloride/acrylonitrile copolymers
compared to weight-loss of polyvinyl chloride.

KINETICS

The rate of weight loss is plotted versus temperature for the
random (R-2) and alternating (A-1) copolymers of 0.38 RV in Fig. 6.
The rate of weight loss was calculated at each point by using convoluted
integers [8]. The use of convoluted integers is equivalent to smoothing
and differentiating by least squares polynomials. For the rates in Fig. 6,
9-point quadratic convoluted integers were used. That is, each rate is
equivalent to fitting a quadratic expression through 9 points (4 points on
each side of the point of interest) and evaluating the derivative of the
quadratic expression at that point.

The maximum rate of weight loss is lower and the decomposition is
distributed over a broader temperature range for the random copolymer than for the alternating copolymer. The difference in the
shape of the weight-loss curve suggests that the mechanism of dehydrohalogenation is different in the random than in the alternating copolymer.
As might be expected from the differences in sequence distribution, the
relative contributions of initiation and propagation to the weight loss are
probably different in the two copolymers.

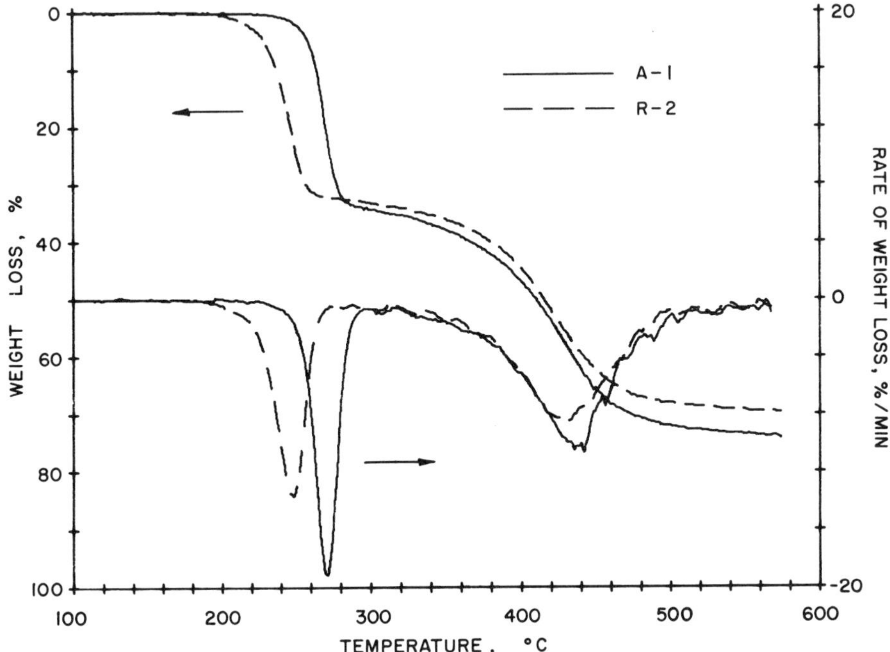

FIG. 6. Rate of weight-loss vs temperature for random (R-2) and alternating (A-1) vinyl chloride/acrylonitrile copolymers.

Figure 7 illustrates that the maximum rate occurs at a lower conversion for the alternating copolymer than for the random copolymer. The conversion was calculated as the ratio of the weight loss at each temperature to the total weight loss attributable to generation of HCl. Assuming that the weight-loss can be described by

$$dW/dt = (dW/dT)r = Ze^{-E/RT}W^n \tag{2}$$

Flynn and Wall [9] found that the maximum rate shifts toward lower conversion as the reaction order, n, increases. Therefore the results in Fig. 7 suggest that the order of the reaction is higher for the alternating copolymer than for the random copolymer. That is, the concentration of some reactant controls the decomposition of the alternating copolymer more strongly than that of the random copolymer. For example, the elimination of HCl might be expected

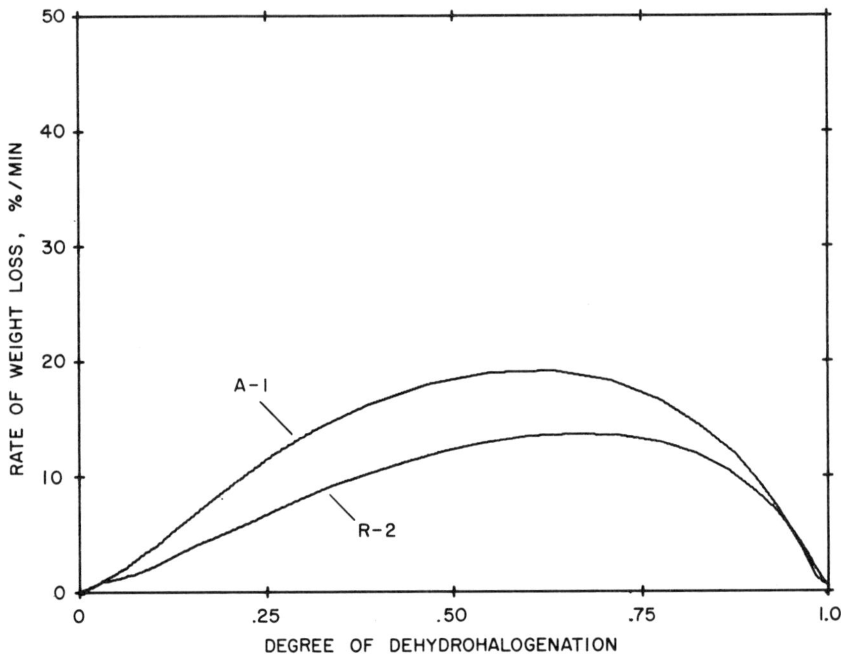

FIG. 7. Rate of weight-loss vs degree of dehydrohalogenation for R-2 and A-1.

to be controlled by the concentration of initiation sites in the alternating copolymer because propagation cannot occur along runs of VCl units.

Figure 8 compares the rates of weight loss calculated from the dynamic data with the rates of weight loss measured at constant temperature near the threshold of weight loss for R-2 and A-1. Even after smoothing the dynamic data to 25 points, there is still considerable scatter in the data at low conversion, so that the dynamic rates and isothermal rates are only approximately in agreement. This illustrates the difficulty in using dynamic data for kinetic studies at low conversions. For the isothermal rates included in Fig. 8, the rates were constant for at least 30 min. Based on the isothermal data, the activation energy which could be calculated from Eq. (2) would be approximately the same for the two copolymers, but the frequency factor would be considerably less for the alternating copolymer than for the random copolymers.

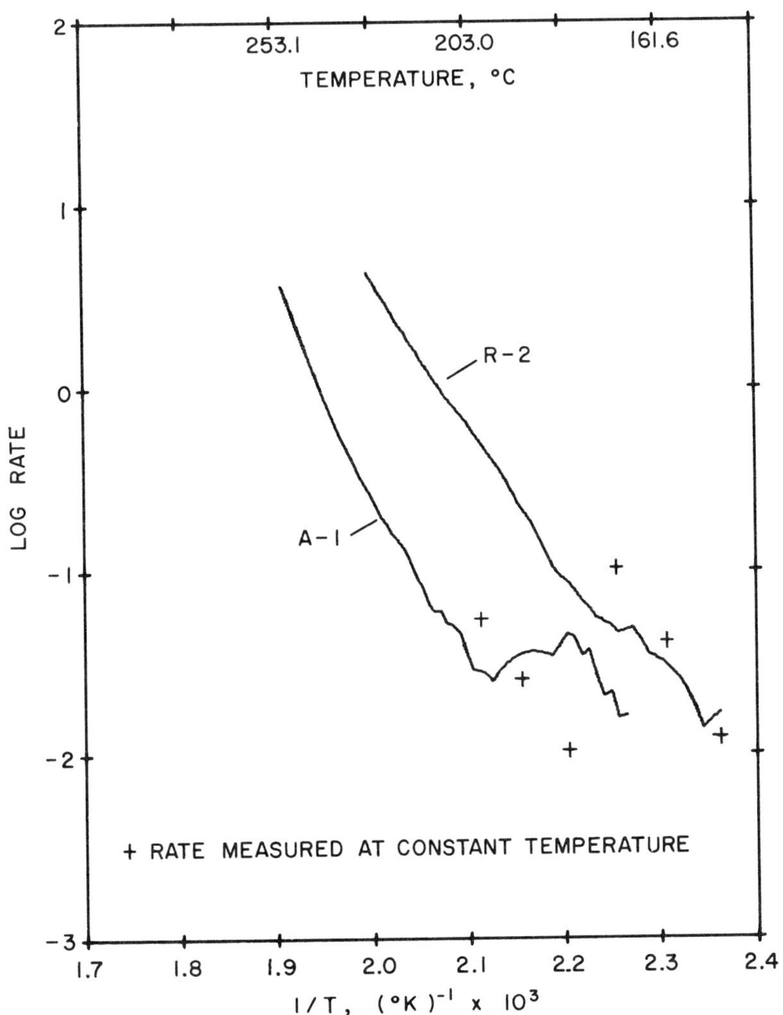

FIG. 8. Logarithm of rate of weight-loss vs reciprocal of absolute temperature for R-2 and A-1.

ACKNOWLEDGMENTS

The authors are grateful to Mr. J. E. Simborski for assistance in preparing the polymers and to Dr. P. W. Kopf for bringing the concept of convoluted integers to their attention.

REFERENCES

[1] N. W. Johnston, Polym. Preprints, 13, 1029 (1972).
[2] G. Wentworth and J. R. Sechrist, J. Appl. Polym. Sci., 16, 1863 (1972).
[3] S. D. Norem, M. J. O'Neill, and A. P. Gray, Thermochim. Acta, 1, 29 (1970).
[4] E. A. Boettner, G. Gall, and B. Weiss, J. Appl. Polym. Sci., 13, 377 (1969).
[5] J. Furukawa, E. Koboyashi, and J. Yamauchi, Polym. J., 2(3), 407 (1971).
[6] H. J. Harwood, N. W. Johnston, and H. Piotrowski, J. Polym. Sci., C, 25, 23 (1968).
[7] W. I. Bengough and H. M. Sharpe, Makromol. Chem., 66, 31 (1963).
[8] A. Savitzky and M. J. E. Golay, Anal. Chem., 36, 1627 (1964).
[9] J. H. Flynn and L. A. Wall, J. Res. Nat. Bur. Stand., A, 70, 487 (1966).

The Use of Derivative Thermogravimetry to Estimate Degree of Thermal Degradation

THOMAS J. GEDEMER

Research and Development
McGraw-Edison Power Systems Division
Franksville, Wisconsin 53126

ABSTRACT

It is difficult to estimate the degree of thermal degradation
which has taken place in complex parts composed of
cellulosics and polymeric insulation unless complete fail-
ure has occurred. There is a definite need for a parameter
which can quickly and easily measure the relative degree
of thermal attack on organic insulation subjected to various
known environmental and processing conditions. This paper
reports preliminary work using derivative thermogravimetry
to perform this function. Derivative thermogravimetric
analysis (TGA) was used to produce a value, TGA Index,
related to the peak height of the major TGA derivative signal.
This TGA Index was found to correlate with the copper
number of insulating paper measured after thermal degradation. The application of TGA Index to evaluation of polymer
degradation after thermal shock, water absorption, and accelerated UV attack is also discussed. Finally, preliminary
studies on the use of derivative TGA to evaluate coatings exposed to various outdoor environments around the country
are reported.

95

INTRODUCTION

Thermogravimetry (TG) or thermogravimetric analysis (TGA) has been used for many years to study degradation processes in polymers. Reaction mechanisms and kinetics have been elucidated in this manner, and accelerated aging tests have been developed which use TGA to predict relative thermal stability of a wide range of materials. Because thermogravimetry is such a versatile tool in the study of degradation processes themselves, it was the logical method to explore evaluation of a related phenomenon, the state of degradation.

The rationale for our initial experiments can be summarized as follows. Thermogravimetry involves the dynamic measurement or weight changes in a sample as a function of temperature. A TGA run on a sample of fresh material will usually exhibit weight loss steps corresponding to various stages in the degradation of the original substance plus weight losses corresponding to secondary reactions of the initial degradation products. At any given temperature T_x, the rate of weight loss is a function of the amount of original material remaining and the nature and relative amounts of degradation products present.

If one now submits a sample of the fresh material to thermal degradation and runs another TGA experiment on the degraded material, the rate of weight loss at T_x is still a function of the amount of original material remaining and the activity of the degradation products. However, we know that there is less of the original material present since some prior breakdown has taken place. If the primary breakdown products of degradation are relatively inert, they will exert a dampening effect on the rate because of dilution effects, and the result will be a slowing of the rate of weight loss. If, however, the major breakdown products are reactive, an increase in their concentration due to prior degradation could very well lead to an increase in the rate of weight loss for the degraded specimen. Either way, a definite change in rate of weight loss should be observed in degraded material.

EXPERIMENTAL

Figure 1 illustrates a typical weight loss curve obtained with organic electrical insulating materials. The most sensitive point at which to choose T_x would be at the maximum rate of weight loss. Unfortunately, in the conventional thermogram, this corresponds to the point at which the curve is changing direction at the greatest rate and reproducible measurements are very difficult to make. However, if one records the time derivative of the weight loss against temperature

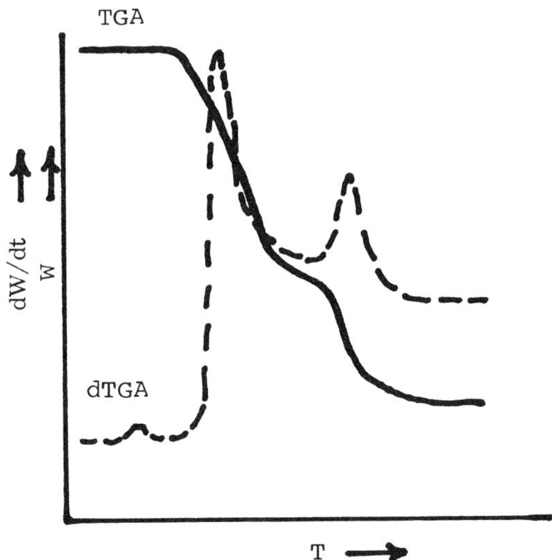

FIG. 1. Typical thermograms of cellulose insulation.

instead of weight loss, a curve similar to the dashed line in Fig. 1
results. The point of maximum rate of weight loss is now displayed
as the maximum of a well-defined peak. This maximum is easily
measured and is quite reproducible.

For this work, a Du Pont Thermal Analyzer with a Du Pont 950
Thermogravimetric Analyzer was used to produce the primary TGA
curve. The output from the thermobalance was processed by a
Cahn Time Derivative Computer and the dTGA curve was then recorded on a Houston X-Y recorder. Sample size was generally
between 2 and 6 mg.

RESULTS

The dashed curve in Fig. 1 illustrates the typical time derivative
weight loss curve obtained with insulating paper. Three maxima
are observed, a small peak at 30 to 40°C, a large peak at 335 to 340°C,
and a third peak of variable size centered at about 440°C.

The derivative traces in Fig. 2 show the basis of the method. After
aging for 16 hr at 200°C, cellulose insulation is degraded to the point
where the major weight loss on thermogravimetry, the one whose
derivative signal peaks out at 340°C, has been greatly retarded. The

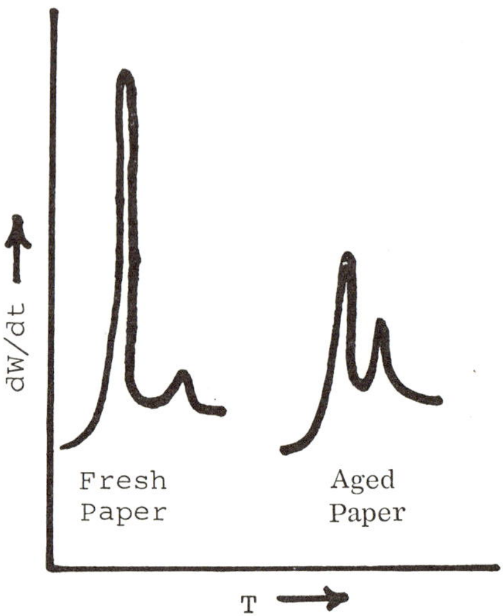

FIG. 2. Comparison of fresh and aged paper (16 hr at 200°C).

quantitative measurement of this rate retardation is what we are relating to prior degradation. The value which we have labeled as TGA
Index is defined as the height of this peak in inches of chart paper
divided by the original sample weight, T.I. = h/w.

Table 1 summarizes the results of a reproducibility study on fresh
paper and pressboard. The individual values recorded in Table 1 were
obtained over a period of several weeks so that the precision calculated
is a reasonable estimate of the precision likely to be encountered over
the time span of a single series of experiments. With both pressboard
and paper the relative standard deviation is less than 10%. This is a
reasonable precision for the type of measurement being made.

The next question to answer is whether these curves obtained by
thermogravimetry correlate with any of the more traditional methods
for estimating thermal degradation in insulation. With cellulose, one
of the tests widely used is copper number.

Copper number has been used for many years to evaluate paper insulation. In theory, copper number denotes the number of scissions
formed by aging. In practice, the reagent reacts more or less
selectively on ketone and aldehyde groups. It is a very useful test,
especially when used in conjunction with acid number.

The bow in the curve of Fig. 3 is caused by the fact that the TGA
Index is a double-valued function with respect to copper number. The
index rises slowly as aging proceeds at low temperatures. This may
represent a small improvement in properties due to loss of loosely

TABLE 1. Reproducibility of TGA Index Values

	h/w Values of 60% manila/40% kraft paper	h/w Values of 100% kraft pressboard
	3.41	3.71
	3.54	3.34
	2.86	3.53
	3.54	3.88
	3.02	3.16
	3.74	3.68
	3.65	3.62
$\overline{X}$	3.39	3.64
S	0.33	0.31
Range	3.06 to 3.72	3.33 to 3.97
$\overline{X}/S$, %	9.7	8.5

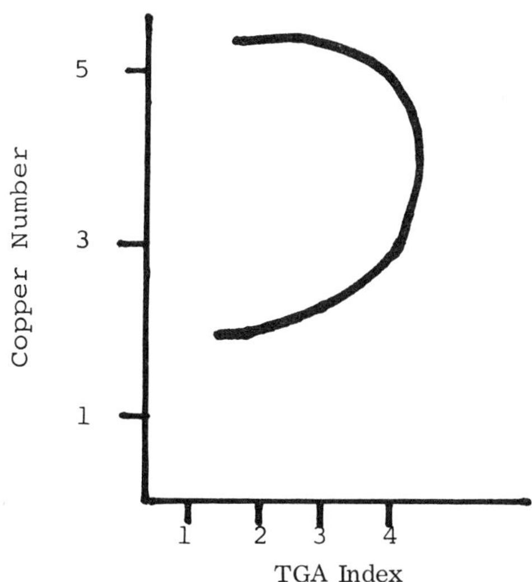

FIG. 3. Copper number vs TGA index for insulating paper.

bound water. As the aging temperature is increased, a point is reached
at which the index rises sharply; this corresponds to the region of
greatest increase in copper number and so it must be considered to be
the point where serious degradation has begun. In contrast to copper
number, which slowly approaches a limiting value of 6 or 7 as degradation proceeds, the TGA Index reaches a maximum and then drops off
very sharply to quite low values for completely charred paper.

Since we were fairly successful in evaluating the thermal degradation
of cellulose by derivative thermogravimetry, we decided to investigate
the utility of this technique with other materials and other modes of degradation. The example we chose was a poly(vinyl chloride) powder
coating we had been evaluating.

We ran derivative thermogravimetry experiments on fresh coating
and coating after exposure to three separate environments. The first
test was a simple immersion in water at 38°C for 14 days. The second
was a thermal cycle test in which the temperature was cycled from -23
to 130°C twice a day. The coating was examined after 10 cycles. The
third test is an UV/condensing moisture test conducted in a QUV cabinet
made by the Q-Panel Co. The coating was examined after 1469 hr of
UV exposure.

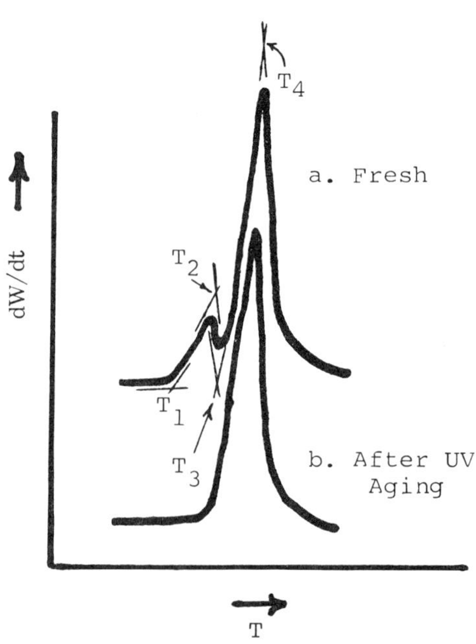

FIG. 4. Comparison of fresh and UV-aged PVC coating.

TABLE 2. DTGA Results on PVC Powder Coating

Treatment	h/w, T_1	h/w, T_2	h/w, T_3	h/w, T_4
Fresh	0.16	0.42	0.51	2.3
After 14 days at 38°C in water	0.18	0.44	0.60	2.2
After 12 cycles of -10° to 266°F	0.08	0.23	0.41	2.4
After 1469 hr UV exposure in QUV cabinet	0.30	-	0.62	1.9

A typical derivative TGA trace of the fresh PVC formulation is shown in Fig. 4. Measurements were made at the four indicated temperatures, which in the case of fresh material were 249, 255, 266, and 278°C, respectively. A summary of the TGA Index values obtained is given in Table 2. The water immersion test revealed no changes in gloss or Sward rocker hardness, and essentially no change in the DTGA results either. No cracking, embrittlement, or loss of adhesion was observed after the thermal cycling test, but the h/w values of T_1 and T_2 dropped to half of the original value, suggesting that some physical change has occurred in the material. After exposure to 1469 hr of UV radiation, one might expect some serious changes in the formulation to have taken place, but all that is observable is a moderate amount of chalking. The DTGA results indicate drastic changes however, the peak associated with T_2 disappearing entirely, as shown in Fig. 4.

One may postulate that the rate processes involved with T_1 and T_2 are associated with more volatile components of the formulation. The lower TGA Index values for T_1 and T_2 observed after thermal cycling can then be explained by assuming the more volatile fragments have been removed at the high temperature end of the cycle. The more drastic changes observed after UV exposure can be related to the degradation of the main PVC polymer itself, which results in a lowered TGA Index for T_4. The volatile breakdown products then provide the rationale for the doubling of the T_1 rate. The disappearance of the T_2 rate maximum may be due either to the complete reaction of a specific formulation component responsible for T_2 or it may be due to the merging of the T_2 maximum with rate maxima produced by primary and secondary degradation products. These postulated mechanisms can be readily checked by further work, but it is clear that the derivative TGA method described here can serve as a sensitive indicator for many types of degradation.

TABLE 3. One Year Outdoor Weathering Results dW/dt

Location	Peak temperature (°C)			Peak height (mV)		
	Alkyd 1	Alkyd 2	Acrylic	Alkyd 1	Alkyd 2	Acrylic
Control	380	334	346	0.63	0.73	0.85
Mild Industrial	374	329	393	0.55	0.70	0.85
Hot dry	354	326	307	0.54	0.55	0.78
Hot moist	317	335	382	0.52	0.50	0.60
Severe industrial	357	324	375	0.47	0.59	0.86

By the same token one would expect the derivative TGA procedure to be quite helpful in evaluating the effects of outdoor weathering. We presently expose organic coatings to four separate locations around the country considered to be, respectively, mild industrial, hot dry, hot moist, and severe industrial environments. A sampling was made of three coatings which had been exposed for 1 year, and Table 3 summarizes the TGA data we obtained from them.

The first point to note is that the two alkyds act similarly, both yielding lower peak heights and lower peak height temperatures after exposure. The fact that the TGA degradation rate drops suggests that the products of weathering breakdown are inert, perhaps even inhibitive, with respect to subsequent thermal degradation. Because the peak temperatures also drop, one can further surmise that the weathering breakdown products are lower molecular weight fragments of the original formulation.

Let us assume that, for alkyd systems at least, the greater the drop in peak height, the greater the weathering degradation. Certainly, such an assumption must be verified by correlation with definitive property data and, moreoever, will depend entirely on the physical property chosen for correlation. But we expect our TGA method to perhaps show degradation trends before they are apparent by other means, and a 1-year exposure is really not enough to expect discernible changes in good coatings by traditional methods. So let us bypass the problem of correlation for now and see what we can extract from the data shown.

If one does make the aforementioned assumption, it is clear that alkyd formulation #1 performs about as expected with mild industrial and hot dry environments affecting the coating the least; the hot, moist environment affecting the coating a little more; and the severe industrial environment causing the most serious degradation. The peak temperature drops show the same trend except that the hot, moist location exhibits a much lower peak temperature than the severe industrial site, which actually shows up better than the hot dry location. A possible explanation for this is that the hot moist location is also located near the ocean and is subjected to a certain amount of salt spray. This undoubtedly leads to the presence of metallic salts on the surface of the coatings, and these salts may well catalyze the TGA degradation reactions.

In contrast, the peak heights for alkyd formulation #2 suggest that the hot, moist temperature is the most degrading. Even the hot, dry environment gave a lower peak height than the severe industrial site. The fact that both warm climate sites exhibited the most attack suggests that alkyd formulation #2 is temperature sensitive, and further tests should be conducted before the coating is ever used in hot climates. The peak temperatures for alkyd formulation #2 show that a slight catalytic effect has occurred at all of the sites except at the hot, moist site. This is in direct contrast to the results from the first alkyd formulation, and while it can be rationalized, any explanation should really await the results of further experimentation.

The third example is an acrylic coating. The attack of either industrial environment on the acrylic is not evident after a year's exposure. Both hot weather sites, on the other hand, show evidence of having affected the acrylic coating, the moist site doing the most damage. The apparent severity of the two warm climate sites may be due to the higher temperatures, but most likely it is the increased incident UV radiation because of the additional sunlight at these sites that is responsible for the attack.

This particular acrylic was chosen for inclusion because it demonstrates anomalous temperature effects. The coating panels exposed at both industrial sites and at the hot, humid site exhibit increased peak temperatures over the control, suggesting that cross-linking may be the primary reaction taking place. However, the panels from the hot, dry site show a considerable drop in peak temperature, suggesting disruption of polymer chains as the major route. Further work is needed to clarify this situation, but the mechanism of acrylic degradation at these outdoor sites is obviously more complex than one would expect at first glance.

CONCLUSIONS

The cellulose and polymer studies reported suggest that the derivative TGA method may be applicable to a wide variety of aging and processing conditions. The advantage of the TGA Index approach is that by tailoring aging conditions to approximate known processes, a simple, fast, effective method for post-mortem estimates of thermal attack on insulation can be developed for the monitoring and evaluation of many plant processes.

Bearing in mind the constraints on this approach, we can draw the following conclusions:

1. TGA Index, as defined above, is a very useful parameter for evaluating, after the fact, the degradation, thermal or otherwise, suffered by polymers.

2. The TGA Index has a precision of 5 to 10% in estimating degree of thermal degradation.

3. For cellulose insulation, TGA Index correlates with copper number in evaluation of thermal degradation.

4. TGA Index shows promise as a fast, sensitive monitor for early detection of degradation under natural and artificial weathering conditions.

The Evaluation of Two Pyrolyzers for the Analysis of Insoluble Polymers

PETER CUKOR and CARMINE PERSIANI

GTE Laboratories Inc.
Waltham, Massachusetts 02154

ABSTRACT

Two devices were evaluated in the pyrolysis-gas chromatographic analysis of a linear polyethylene standard. The Chemical Data Systems Coil Probe, a pulse-type pyrolyzer, and the Du Pont 950 Thermogravimetric Analyzer, a continuous heat-type unit, were studied. The latter was connected to the gas chromatograph using an interface designed to allow on-column collection of pyrolyzates.

It was found that the coil probe yields reproducible results in the analysis of linear polyethylene, and the distribution of pyrolysis products obtained is essentially unaffected by changes in heating rate, time, and final heating temperature, provided these parameters are above certain experimentally determined values. This behavior may be explained by the action of the quartz tube sample holder as a heat transfer barrier between the heating coil and the sample. Large size samples enhance the formation of secondary products. The most serious shortcoming of the device is the difficulty involved in quantitating sample size.

It was further found that the proposed thermogravimetric analysis-gas chromatography interface had only limited usefulness because of back-diffusion. Condensation in the

transport lines and formation of secondary products presented
some problems. The results were reproducible and product
distribution was not effected by changing heating rate and
sample size. Isothermal operation of the thermal balance
yielded satisfactory results and shortened the analysis time
considerably.

INTRODUCTION

In order to further the art of pyrolysis-gas chromatographic analysis of insoluble polymers, we have evaluated the performance of two
pyrolyzers, the Chemical Data Systems Coil Probe, a pulse-type unit,
and the Du Pont 950 TGA, a continuous heat pyrolyzer. Both of these
instruments are capable of handling genuine solid samples without the
use of solution techniques, and they also allow controls over such
important variables as heating rate, temperature of pyrolysis and
heating time. In the course of this work, a new thermogravimetric
analysis-gas chromatography (TGA-GC) interface utilizing on-column
trapping rather than external traps was used and evaluated.

The technique of pyrolysis-gas chromatography has been reviewed
extensively [1-3]. This method of analysis has not developed to its
full potential because of lack of interlaboratory communication due to
the use of nonstandardized equipment and operating methods.

The first generation of pyrolyzers were either the oven [4] or the
filament type [5]. The only variables which were controlled to some
extent were the final temperature and time of pyrolysis. Consequently,
the technique was marked by irreproducible results. The search for
an optimized pyrolyzer with improved design and operation resulted
in the recognition that the method of heating, the heating rate, and the
sample size and geometry are important parameters to be considered
[2, 6]. A second generation of pyrolyzers was developed and
characterized by pulse-type application of heat as opposed to the continuous method of heating of the oven-type pyrolyzers [2]. The pulse-
type pyrolyzers are characterized by extremely rapid heating rate
(rise time), accurate control of final temperature, and extremely
small sample size. The Curie point pyrolyzer [7, 8] and the Chemical
Data System ribbon probe [9] are examples of pulse-type pyrolyzers.
It is required in the operation of these instruments that the polymers
investigated be dissolved in a solvent. The aliquots of this solution are
then transferred onto the pyrolyzer where, upon evaporation of the
solvent, a polymer film remains which is subsequently pyrolyzed. The
technique described above is obviously not applicable for cross-linked
and/or insoluble polymers, but these are the materials in whose
characterization pyrolysis-gas chromatographic analysis is most
essential.

EXPERIMENTAL

Reagents

Linear polyethylene:	NBS Standard 1475
Polystyrene M_w = 2,500,000:	Waters Associates
Polystyrene M_w = 5,000:	Waters Associates
Hydrocarbon kits:	Polyscience Corp.

Apparatus

Chemical Data System Pyroprobe 100 equipped with ribbon and coil probe. The latter was used exclusively in the present work. It consists of a platinum coil into which a quartz tube containing the sample is placed. It is possible to control the heating rate, the final temperature of the coil, and the time of heating.

Du Pont 950 Thermogravimetric Analyzer modified to allow purge gas to enter at the back (tare weight end) of the instrument.

Perkin-Elmer 900 Gas Chromatograph equipped with dual columns and dual flame ionization detectors and without a subambient over cooling accessory.

The interface between the thermal balance and the gas chromatograph was designed for on-column collection of pyrolyzates with the use of cold traps. A Pyrex female ball joint was attached to the quartz furnace tube of the TGA with a clamp. The opposite end of the Pyrex tube was connected through a graded seal to 1/16 in. o.d. stainless steel tubing, which in turn was attached via a "swage lock" fitting to the injector insert of a Perkin-Elmer gas sampling valve. The exit portion of the furnace tube which is outside the oven and the adjoining transport lines were maintained at a temperature of $250°C$ with the aid of Perkin-Elmer "mass spec"-type heating tapes and variacs. Two separate tanks of helium are required: one serves as carrier gas for the GC while the other is used as the purge gas for the TGA. A schematic diagram of the TGA-GC system is shown on Fig. 1.

Procedure

CDS Coil Probe

About 500 μg of polymeric material was placed in the quartz tube which was then inserted into the platinum coil of the probe. The assembly was attached to the gas chromatograph and allowed to equilibrate for 10 min, at which point the sample was pyrolyzed and analyzed by the gas chromatograph.

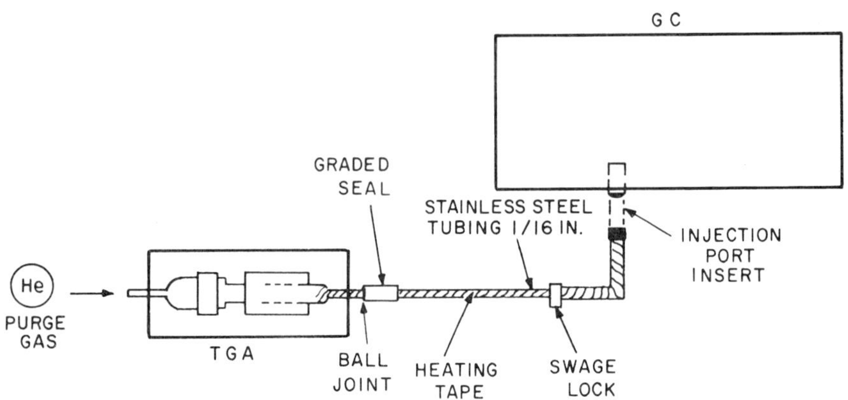

FIG. 1. Schematic representation of TGA-GC interface.

Thermogravimetric Analyzer

The sample was placed into the instrument which was then purged
with 200 ml/min of helium for 30 min without being connected to the
GC. Simulatneously, the gas chromatographic column was cooled to
0°C. The helium flow to the gas chromatograph was then shut off,
the septum cap removed, and the injector insert of the interface was
placed into the gas chromatograph. At this point the gas flow through
the columns was supplied by the TGA purge gas. The sample was then
heated to 900°C and the pyrolyzate collected on the GC column. Upon
completion of the pyrolysis step, the injector insert was removed from
the chromatograph, the septum cap was replaced, and the carrier gas
flow reinstated. After the establishment of equilibrium, the pyrolyzate
was analyzed gas chromatographically.

Gas Chromatograph

The operating conditions were as follows:

Column:	12 ft, 1/8 in. diameter 10% Dexsil on Chromosorb W
Detector:	Dual flame ionization
Injector port temperature:	250°C
Manifold temperature:	250°C
Column temperature:	Coil probe experiments: program 50 to 300°C at 13°C/min TGA experiments: program 0 to 300°C at 13 or 26°C/min

Qualitative identification of peaks was achieved through the use of n-alkane mixture calibration curves. Quantitative peak area measurements were done by multiplying the peak height by one-half of the peak width.

RESULTS AND DISCUSSION

Evaluation of CDS Coil Probe

The Chemical Data System Pyroprobe 100 [9] is a versatile instrument with a capability of controlling heating rate, final temperature, and time of heating. It is equipped with two probes. In one, a platinum ribbon is used as a sample holder, heating element, and sensing device, while in the other probe the sample is placed into a quartz tube which in turn is inserted into a platinum heating and sensing coil. The sophisticated controls available on this instrument apply mainly to the operation of the ribbon probe. In the case of the coil probe, the heating parameters of the platinum coil are well controlled, but the quartz tube sample holder represents a formidable barrier to heat transfer between the heating coil and the sample. In spite of this basic shortcoming, the coil probe was found to be a more practical device for analysis of solid samples. The ribbon probe requires that the sample be dissolved and plated out on it or be melted in the solid state. Many polymers are insoluble or decompose upon melting, thus making the use of the ribbon probe impossible. Furthermore, many polymers yield inorganic residues upon pyrolysis which may short out the platinum ribbon.

The variables studied in connection with the coil probe were heating rate, heating time, final heating temperature, and sample size. Linear polyethylene was used as a sample. A typical chromatogram is shown in Fig. 2. Experiments were carried out to study the effect of final temperature on the pyrolysis process. Samples weighing approximately 500 μg were pyrolyzed at 500, 600, 700, 800, or 900°C (these are temperatures of the platinum coil, not of the sample). The heating rate was maintained at 1°C/msec and the heating time was 10 sec. No pyrogram was obtained at 500 and 600°C settings. The distribution of pyrolysis products was quite similar and reproducible at 700, 800, and 900°C settings (Fig. 3). These results are not surprising, for the TGA analysis of polyethylene showed a single sharp decomposition step at 500°C.

Heating rates of 0.1, 1, and 10°C/msec were investigated. In these experiments the final pyrolysis temperature was 900°C and the heating time was 20 sec for 0.1°C/msec and 10 sec for the other two heating rates. Again no marked differences were observed in product distribution as a result of varying the heating rate. Similarly, the product

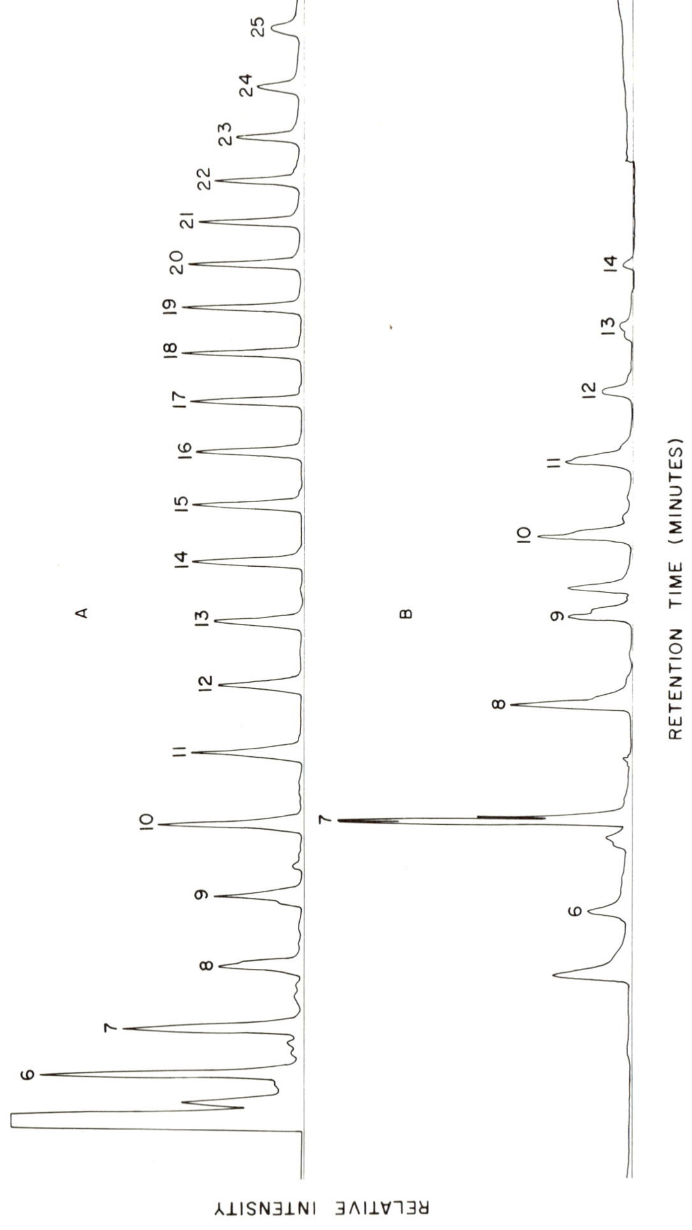

FIG. 2. Typical pyrogram of linear polyethylene using CDS coil probe (A) and TGA-GC system (B).

distribution was not seriously affected by varying the heating time between 2 to 10 sec while maintaining the heating rate at $1°C/msec$ and the final temperature at $900°C$.

Increasing the sample size from 0.5 to 5 mg resulted in the formation of secondary products as shown by the appearance of many extraneous peaks, especially at the low molecular weight section of the chromatogram.

Apparently in the pyrolysis of linear polyethylene with the coil probe, there is a definite temperature ($700°C$ setting) and a heating time (heating time equals rise time) below which pyrolysis will not occur. Above these boundary conditions the behavior of the sample is essentially unaffected by variation in instrument settings. This situation is brought about by the presence of the quartz tube between the heating coil and the sample, and because of the relative insensitivity of polyethylene to changes in pyrolysis temperature. In the analysis of polystyrene the expected change occurred in product distribution as a function of temperature. At $500°C$ the products were styrene and its dimer, while at $900°C$ a great many additional products were also found.

Obviously, reproducibility of results requires careful quality control of the material and geometry of the quartz tube. An additional factor is that upon repeated use, the platinum coil anneals or deforms and will not heat uniformly. According to the manufacturer, this situation has been remedied on the newer models. The greatest single shortcoming of the coil probe is the difficulty involved in quantitating sample size. It requires a microbalance and tedious operation. A $500\text{-}\mu g$ sample size was found to be too great, for a condensed film of polymer was often found at the exit end of the quartz tube.

In order to alleviate some of the problems of quantitization, we are presently involved in the development of an internal standard method for the coil probe. Polymethyl methacrylate upon pyrolysis yields a single major compound, the monomer, and thus can be used as an internal standard. The feasibility of this method is illustrated by the data appearing in Table 1.

Thermogravimetric Analyzer

The application of the thermogravimetric analyzer as a gas chromatographic pyrolyzer was pioneered by Chiu [10, 11]. He has pointed out the advantages of this approach as precise control over heating rate, good reproducibility, no necessity to search for optimized pyrolysis conditions, the possibility of analyzing the products of pyrolysis from each thermal degradation step separately, and the obtaining of important additional information from the thermogram regarding sample size and degradation temperatures. Subsequently, Cukor and Lanning demonstrated the usefulness of an instrument combination

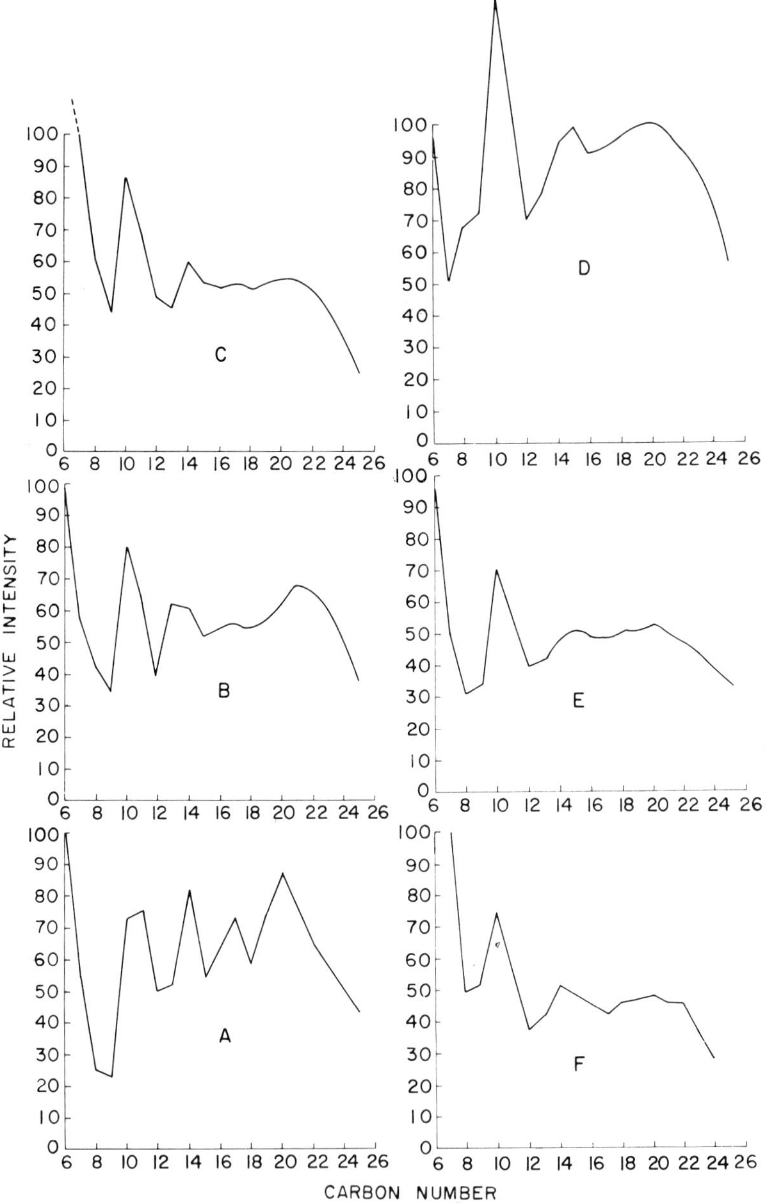

FIG. 3. Relative distribution of pyrolysis products of linear polyethylene using CDS coil probe: continued—

consisting of TGA-GC-IR for the analysis of organic mixtures [12].
Similarly, Chang and Mead combined TGA, GC, and mass spectroscopy
for the analysis of polymers [13]. The present effort is aimed at
continuing these developments. It is directed toward two major areas:
1) investigating the feasibility of on-column trapping with a TGA-GC
system, and 2) evaluating the effect of heating rate and sample size on
the performance of the TGA as a pyrolyzer.

In order to investigate the feasibility of on-column collection of the
pyrolyzate, the effect of prolonged cooling and slow trapping was
initially studied. A mixture of n-alkanes was chromatographed under
several conditions:

1. Injection: temperature programming from 50 to 300°C at 26°C/
   min.
2. Injection: temperature programming from 0 to 300°C at 26°C/
   min.
3. Injection: hold at 0°C for 30 min, then programming to 300°C
   at 26°C/min.
4. Using the TGA for sample introduction employing 30°C/min
   heating rate, collecting on the column at 0°C, then programming
   to 300°C at 26°C/min.

The data obtained indicated that, with the exception of the lowest
molecular weight components of the mixture (C_6 to C_9), no drastic band
broadening was taking place at this rapid heating rate (26°C/min). The
observed band broadening of the low molecular weight hydrocarbons
could probably be eliminated by further lowering the sample collection
temperature. It was also noted that repeated cycling of the Dexsil
column between 0 and 300°C did not affect its performance adversely.

In the course of these experiments, the formation of condensate
films was observed on the wall of the furnace tube toward the cold end

continued from opposite page

Curve	Final temperature (°C)	Heating time (sec)	Heating rate (°C/msec)
A	700	10	1
B	800	10	1
C	900	10	1
D	900	2	1
E	900	20	0.1
F	900	10	10

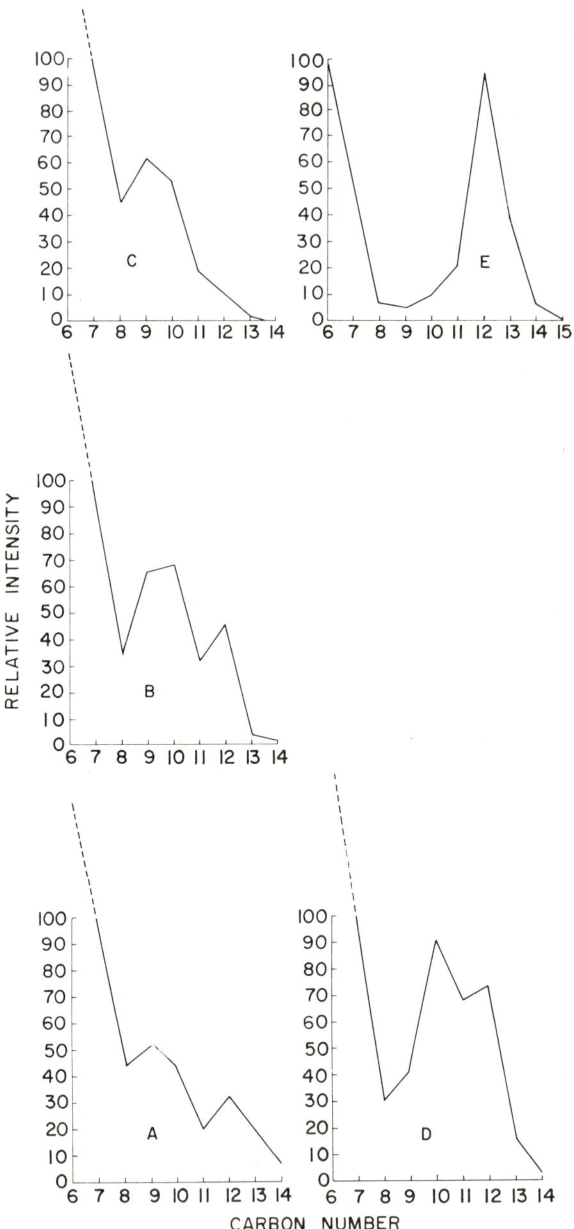

FIG. 4. Relative distribution of pyrolysis products of linear polyethylene using TGA-GC system: continued—

of the balance. The condensation was due to back-diffusion and sputtering of the pyrolyzed fragments. The geometry of the furnace tube, because of its large volume, is not ideal for use as a pyrolyzer. Rapid purging may alleviate this difficulty, but a flow rate of over 200 ml/min was found to cause a large noise signal in the operation of the balance. This back-diffusion problem is particularly serious if the thermal degradation yields gaseous corrosive products such as hydrochloric or acetic acid. In such an instance, severe damage to the balance mechanism will result.

Ideally, transport lines should be maintained at temperatures above the boiling point of the least volatile component. In the case of the pyrolysis of linear polyethylene, this would require maintaining the lines at 400°C, which is a formidable task. Instead, we chose to keep the transport lines at 250°C and allowed hydrocarbons heavier than octadecane to condense at the exit of the furnace tube, making sure all along that this condensate did not produce a memory effect.

Finally, we also noted that secondary reactions occur among the products of pyrolysis. In the pyrolysis of polystyrene, we obtained seven or eight major products instead of the anticipated styrene and its dimer. Chang and Mead reported similar results [13].

The effect of heating rate and sample size on the distribution of pyrolysis products was also investigated using linear polyethylene samples (Fig. 2). No drastic differences in the distribution of products were observed due to varying the heating rate from 10 to 30°C/min or varying sample size from 1 to 25 mg (Fig. 4). A minor difference in product distribution was observed when the TGA had been used in an isothermal manner at 800°C. It took 2 to 3 min for the sample to reach this temperature. The isothermal heating mode shortens the time of analysis considerably, and it is fairly reproducible. The data are summarized in Fig. 4.

---

continued from opposite page

Curve	Heating rate (°C/min)	Sample size (mg)
A	30	1.3
B	30	5.0
C	30	25.0
D	10	4.0
E	Isothermal at 800°C	4.2

TABLE 1. Evaluation of Internal Standard Method

| | Polyethylene | | | Polymethyl methacrylate | Ratios | |
| | | | | | C_{12} Peak/sample wt, mg | C_{20} Peak/sample wt, mg |
μg	Response of C_{12} fragment	Response of C_{20} fragment	μg	Methacrylate response	Methacrylate peak	Methacrylate peak
450	123	102	2	266	1.0	0.85
390	376	228	2	696	1.4	0.84
350	204	125	2	396	1.4	0.91
650	200.0	179	2	286	1.1	0.95

REFERENCES

[1] R. L. Levy, Chromatogr. Rev., 8, 48 (1966).
[2] R. L. Levy, J. Gas Chromatogr., 5, 107 (1967).
[3] B. Groten, Gas Effluent Analysis (W. Lodding, ed.), Dekker, New York, 1967.
[4] K. Ettre and P. F. Varadi, Anal. Chem., 35, 69 (1963).
[5] A. Barlow, R. S. Lehre, and J. C. Robb, Polymer, 2, 27 (1961).
[6] G. Guiochon, Anal. Chem., 40, 998 (1968).
[7] W. Simon, P. Krienler, J. A. Voellmin, and H. J. Steiner, J. Gas Chromatogr., 5, 53 (1967).
[8] J. Q. Walker and C. J. Wolf, J. Chromatogr. Sci., 8, 514 (1970).
[9] A. J. Martin et al., Pittsburgh Conference on Analytical Chemistry and Mass Spectroscopy, Cleveland, Ohio, 1971.
[10] J. Chiu, Anal. Chem., 40, 1516 (1968).
[11] J. Chiu, Thermochim. Acta, 1, 231 (1970).
[12] P. Cukor and E. W. Lanning, J. Chromatogr. Sci., 9, 487 (1971).
[13] T. L. Chang and T. E. Mead, Anal. Chem., 43, 534 (1971).

Degradation Kinetics and Mechanisms of Polymers Using Thermal Spectrometric Techniques

F. ZITOMER and A. H. DiEDWARDO

Analytical Chemistry Department
Celanese Research Company
Summit, New Jersey 07901

ABSTRACT

Techniques utilizing mass spectrometry, thermogravimetry, and gas chromatography in combination to study polymer degradation kinetics and mechanisms are described. The ability to observe volatile species generated under controlled thermal and environmental conditions and the simultaneous accumulation of weight loss data enables acquisition of both kinetic and mechanistic parameters in single run experiments. Polymer degradation mechanisms are best understood when initial oxidative sites can be established. Consequently, data points obtained with these techniques in conjunction with bulk phase analyses provide powerful tools for determining the primary reactions occurring during polymer oxidations.

INTRODUCTION

On-line monitoring of thermogravimetric analyzer effluents by mass spectrometry and gas chromatography provides a means for the

119

continuous analysis of polymer degradation products. Through identification of the volatiles generated and their order of evolution, mechanistic and kinetic data can be readily obtained. In conjunction with bulk phase analyses of the nonvolatile products, the degradation process can be elucidated. Three polymer systems, polymethylene sulfide, polystyrene, and poly-4-methylpentene, serve to illustrate how the technique can be applied.

EXPERIMENTAL

The thermogravimetric-mass spectrometric (TGA-MS) and thermogravimetric-gas chromatographic-mass spectrometric (TGA-GC-MS) analyses were run using procedures and instrumentation described previously [1-7]. All IR spectra were recorded with the Perkin-Elmer 521 grating spectrophotometer on compression molded films of from 2 to 5 mils thickness.

RESULTS AND DISCUSSION

Oxidation of Poly-4-methylpentene

Poly-4-methylpentene is a crystalline, saturated polyhydrocarbon with pendant isobutyl groups on alternate carbon atoms along the polymer chain. It is an interesting polymer structure because there are two potential oxidative sites, the tertiary carbon on the polymer backbone and the tertiary carbon in the isobutyl group. From structural considerations it would be difficult to predict which of the two potential sites would be more susceptible to oxidation because their electron environments are very similar.

Figure 1 shows the IR spectrum of pure isotactic poly-4-methylpentene. It can be seen that the spectrum is free of absorptions in the regions of interest, 1900 to 1600 cm^{-1}. Figure 2 shows the IR spectrum of a sample which has been degraded oxidatively for 2 hr at 190°C.

Four major differences can be observed:

1. The appearance of a band at 3500 cm^{-1} due to hydroxyl terminated linkages.
2. A triplet formation in the carbonyl stretching region, the principal carbonyl absorption being ketonic with underlying absorptions due to peracid, perester, and/or other carbonyl species.
3. A band ingrowth at 888 cm^{-1} due to substituted vinylidine double bonds which form concurrently with the carbonyl groups.
4. A general broadening of the spectrum, especially in the long wavelength regions, which is due to the overall degradation process and oxidative cross-linking.

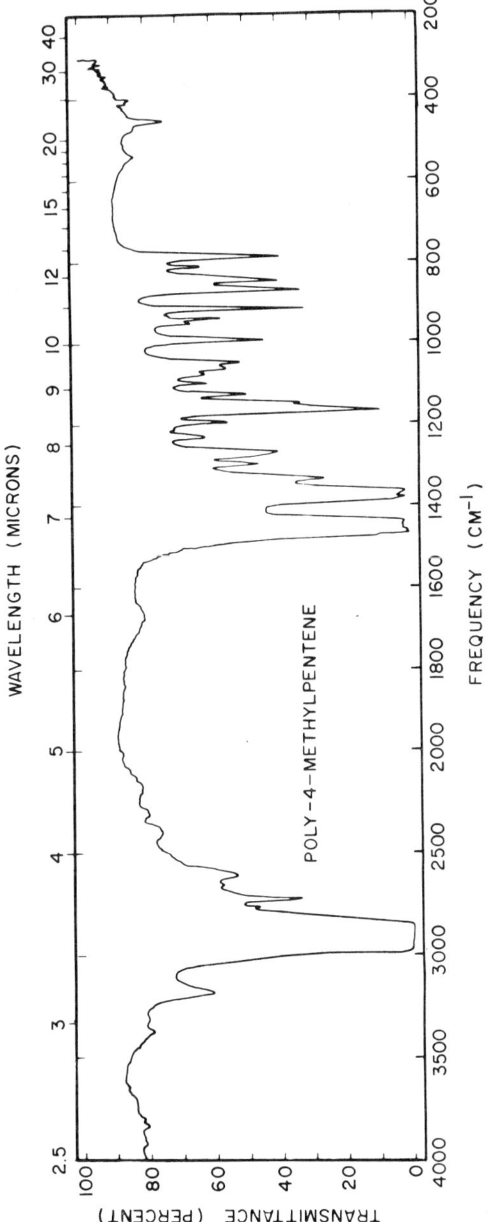

FIG. 1. IR spectrum of poly-4-methylpentene.

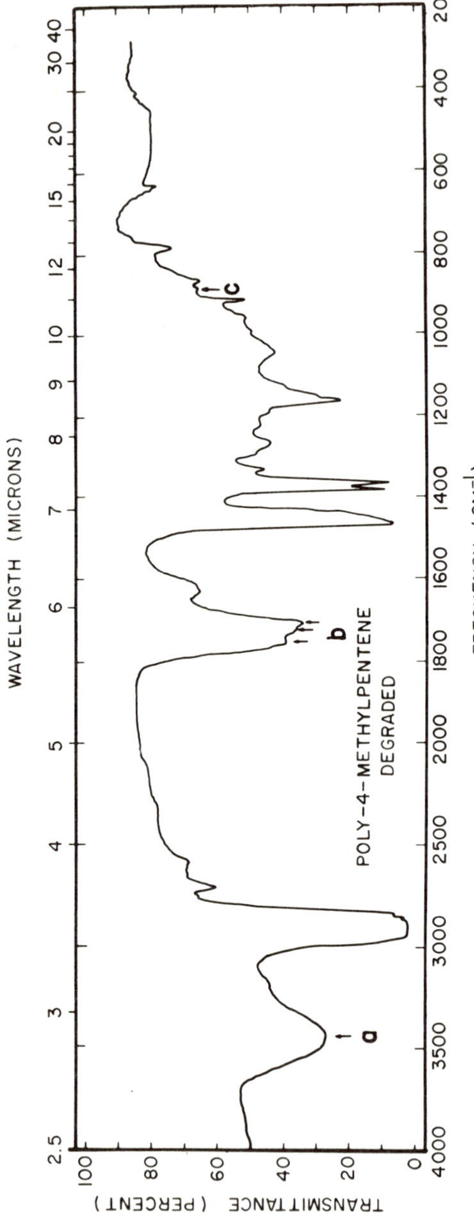

FIG. 2. IR spectrum of poly-4-methylpentene oxidized.

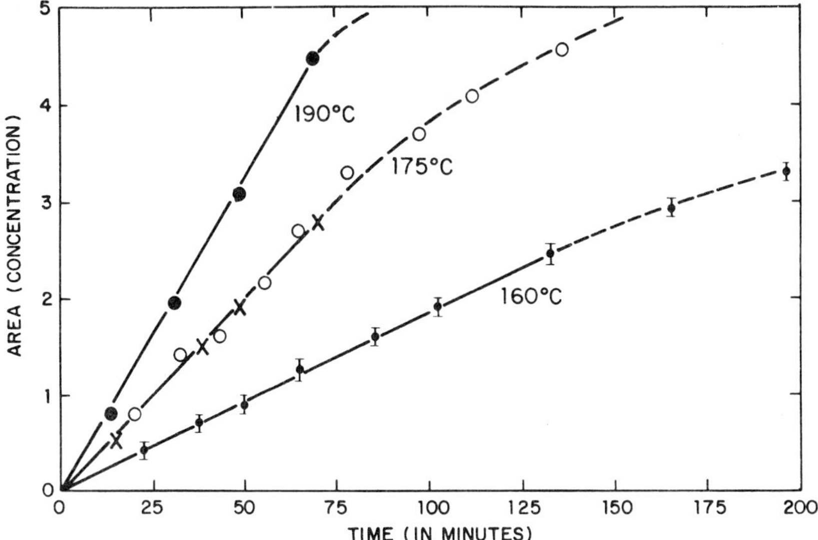

FIG. 3. Rate of total carbonyl ingrowth for poly-4-methylpentene (160-190°C).

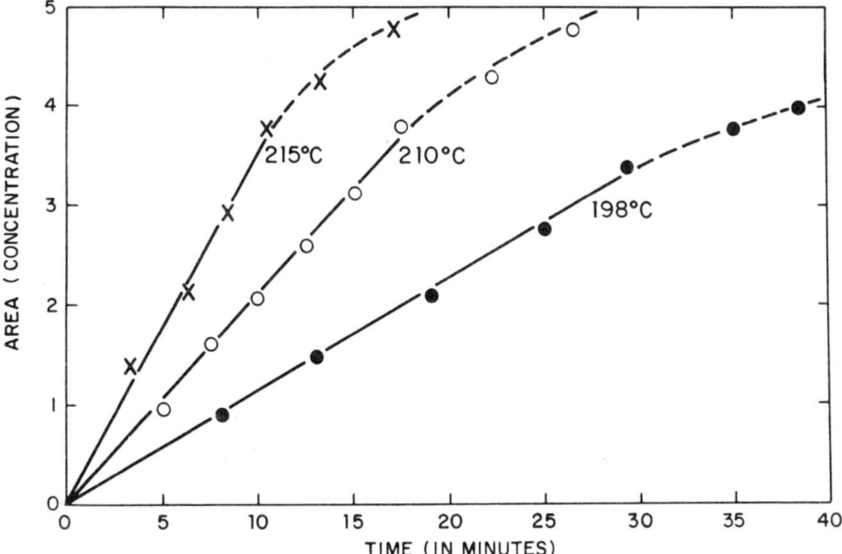

FIG. 4. Rate of total carbonyl ingrowth for poly-4-methylpentene (198-215°C).

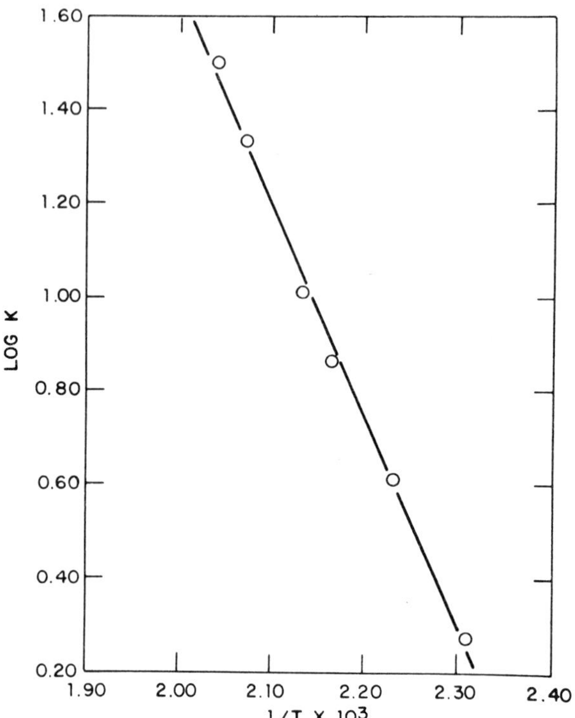

FIG. 5. Arrhenius plot from carbonyl ingrowth of poly-4-methyl-pentene. E_A = 20.1 kcal/mole.

From IR data it is possible to derive the kinetics of the degradation by measuring the rate of carbonyl ingrowth as a function of temperature because the sample thickness and molar absorptivity remain constant throughout the experiment. Consequently, the absorption band area is a direct measure of the concentration of carbonyl species. These data are shown in Figs. 3 and 4. An Arrhenius plot from which the activation energy for carbonyl formation was found to be 20.1 kcal/mole is shown in Fig. 5.

The mechanism of the degradation cannot be derived from this IR data because both oxidative sites would be expected to yield carbonyl and double bond structures subsequent to oxygen attack. Utilizing TGA-MS and TGA-GC-MS data, it is possible to determine both the kinetics and mechanisms of degradation. Fig. 6 shows the thermograms for isothermal degradations of this polymer in air between 160 and

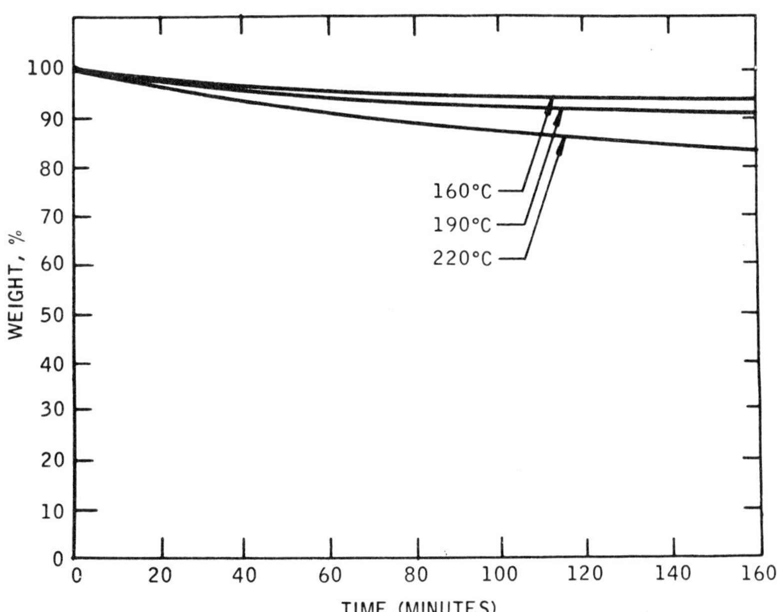

FIG. 6. Isothermal oxidative thermograms of poly-4-methylpentene.

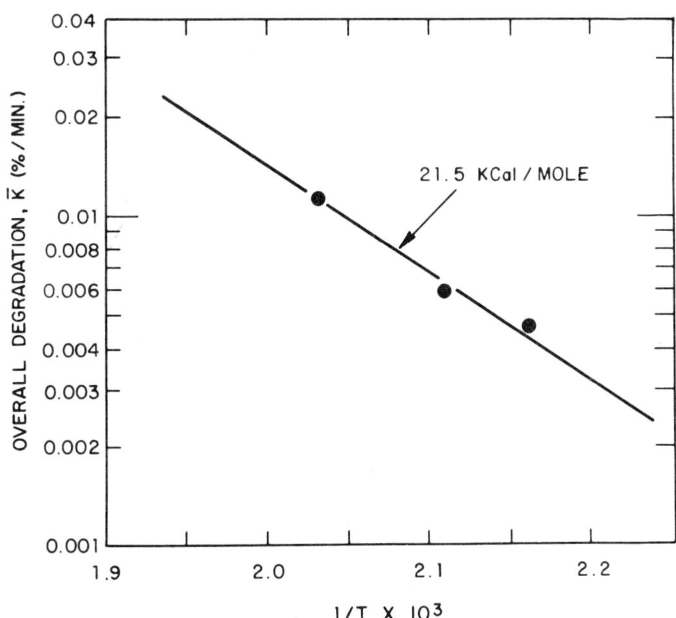

FIG. 7. Arrhenius plot from isothermal oxidative thermograms for poly-4-methylpentene.

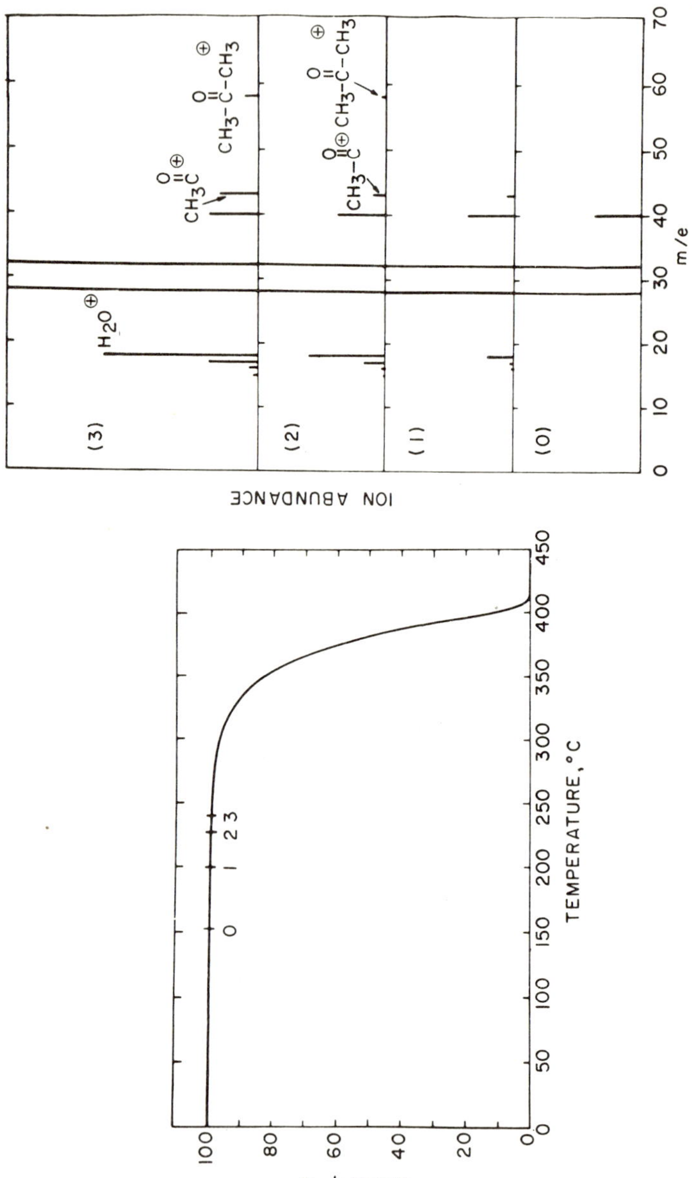

FIG. 8. TGA–MS analysis of poly–4–methylpentene in air.

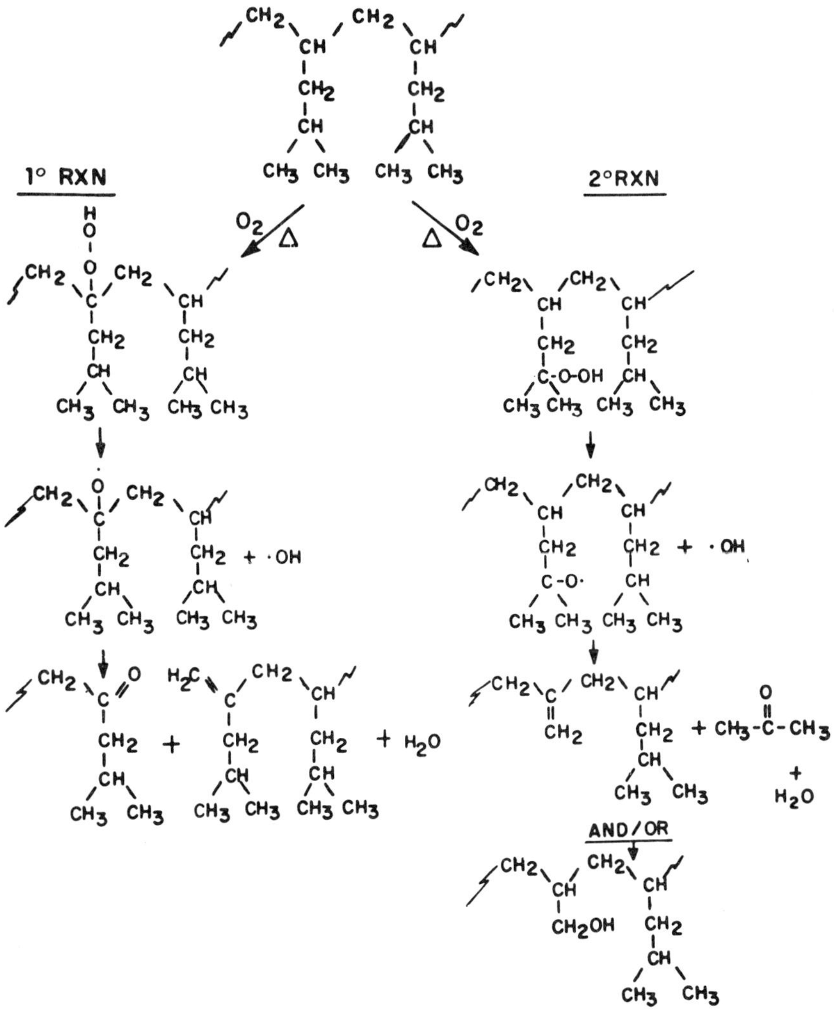

FIG. 9. Proposed mechanism for oxidative degradation of poly-4-methylpentene.

220°C. An Arrhenius plot constructed from this thermal data is shown in Fig. 7 where the activation energy for overall degradation is found to be 21.5 kcal/mole.

TGA-MS data are shown in Fig. 8. Mass spectra were recorded at Points 0 to 3 of the thermogram. Point 0 is the spectrum recorded

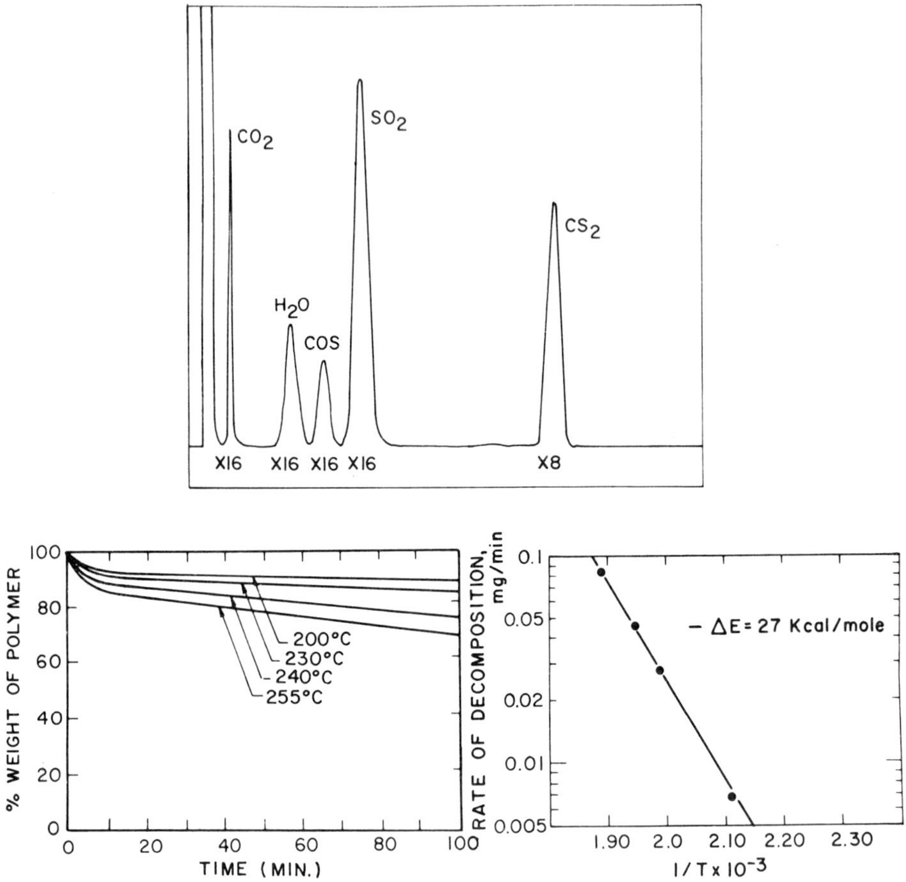

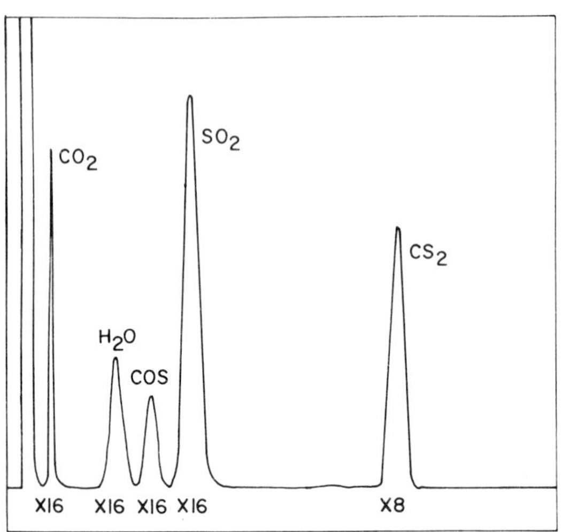

FIG. 10. Thermogram, Arrhenius plot, and chromatogram from thermal oxidation of polymethylene sulfide.

TABLE 1. Quantitative Analysis of Off-Gases

Component	Relative mole %
SO_2	20
CS_2	8
COS	13
H_2O	42
CO_2	17

prior to degradation at 140°C and shows mass peaks associated primarily with the air background. The mass spectrum recorded at Point 1 corresponds to 200°C and shows, in addition to the background spectrum, peaks at m/e 18 and 43 due to the molecular ion of water and the acetyl ion from the fragmentation of acetone. Previous calibration established that the sensitivities of the two peaks are nearly identical. Hence, from the relative peak heights, a 3:1 mole ratio of water to acetone can be approximated. It is interesting to note that Spectrum 1 was obtained before a detectable weight loss was observed

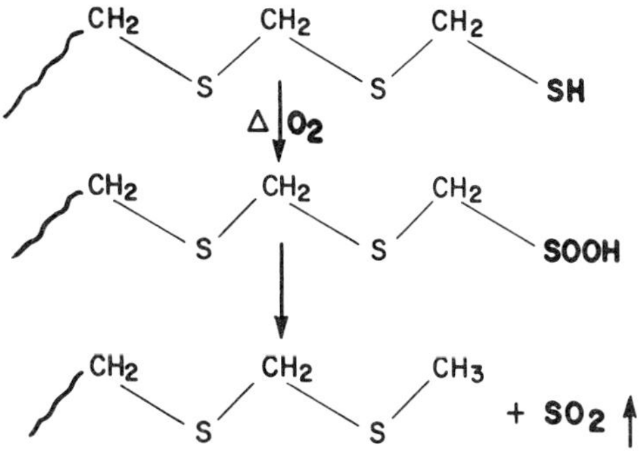

FIG. 11. Proposed mechanism for formation of sulfur dioxide through oxidation of —SH end group in polymethylenesulfide.

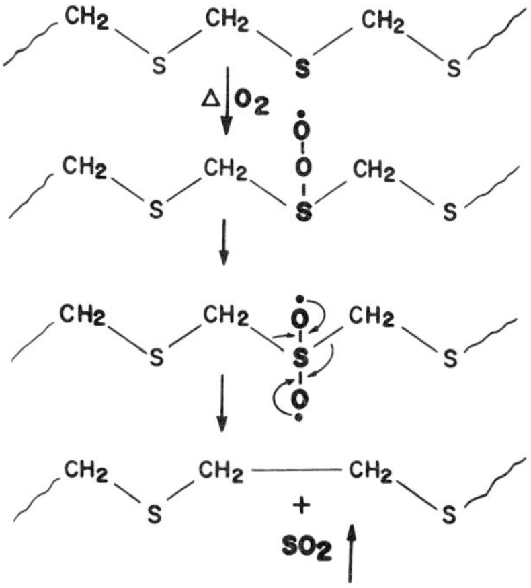

FIG. 12. Proposed mechanism for formation of sulfur dioxide through oxidation of sulfur atoms in polymethylenesulfide.

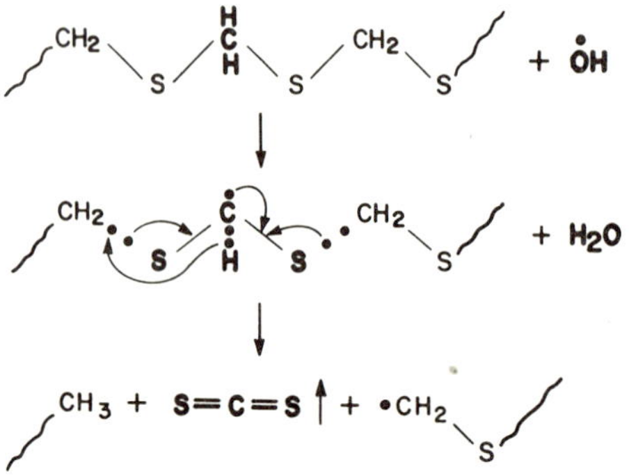

FIG. 13. Proposed mechanism for formation of carbon disulfide and water from degradation of polymethylene sulfide.

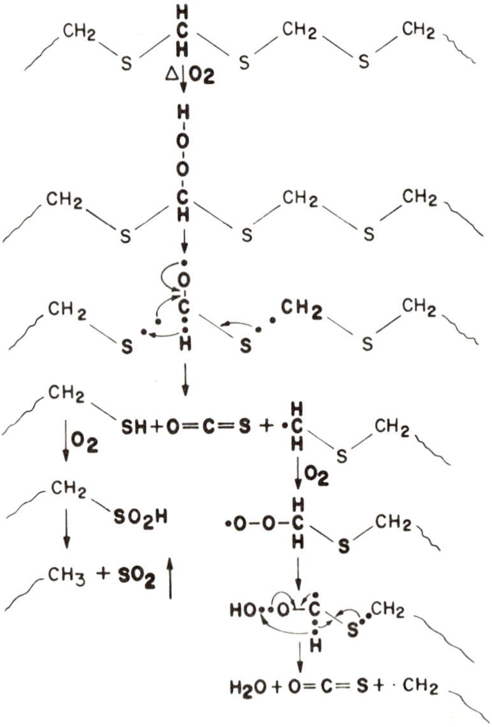

FIG. 14. Proposed mechanism for formation of carbonyl sulfide from degradation of polymethylene sulfide.

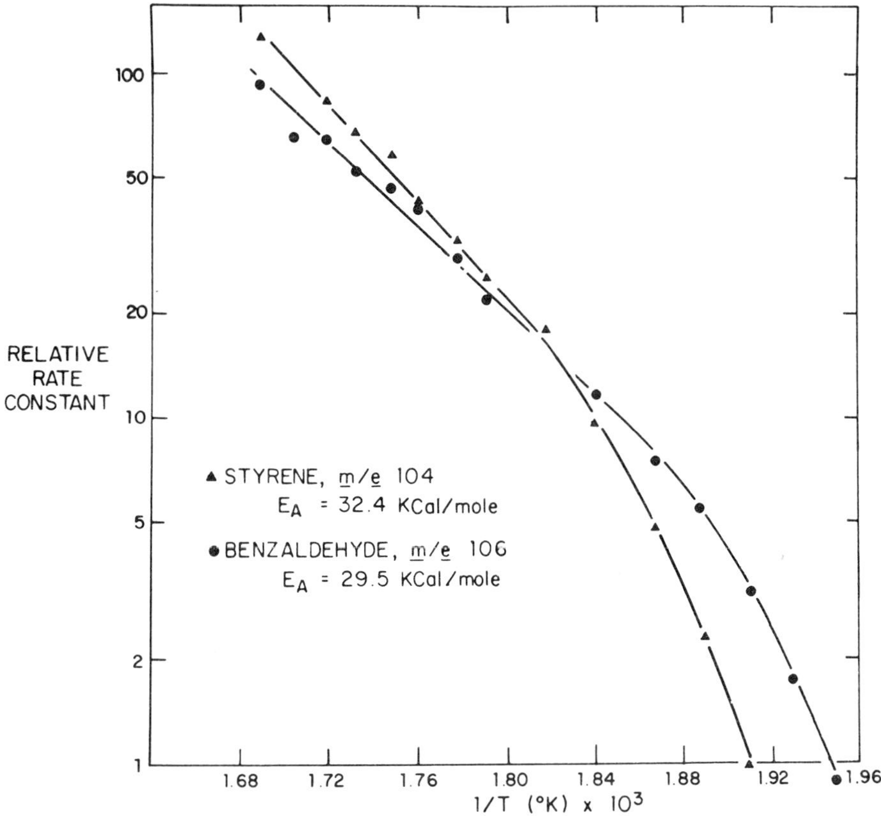

FIG. 15. Rate of formation of styrene and benzaldehyde from degradation of polystyrene.

on the thermogram. This is a good example of the rapid response and high sensitivity of the TGA-MS technique. Spectra 2 and 3, taken at 230 and 240°C, respectively, show the same 3:1 mole ratio of water to acetone. These two spectra also show the molecular ion of acetone at m/e 58 which is much less intense than the acetyl ion. During the later stages of oxidation, water, acetone, and some low molecular weight hydrocarbon fragments were evolded.

A TGA-GC-MS experiment to determine the quantitative ratio of water to acetone showed the ratio to be 3.2:1. This indicates that the relative susceptability to oxidation of the tertiary carbon on the polymer backbone is 2.2 times greater than that of the tertiary carbon on the pendant isobutyl group. Based on these data, the proposed mechanisms for the degradation are shown in Fig. 9.

Polymethylene Sulfide

Polymethylene sulfide (PMS) is an interesting example of how these techniques can be utilized to supplement limited thermo-oxidative data obtained from other spectroscopic methods. IR spectra of PMS films oxidized for varying periods of time showed no significant changes in polymer structure during the initial stages of degradation. TGA-MS and TGA-GC-MS analysis of the degradation products yielded sufficient information to describe the thermo-oxidative behavior. Figure 10 shows an Arrhenius plot for the oxidation degradation kinetics derived from isothermal thermograms. TGA-MS analysis of the off-gases identified the major volatile components as water and sulfur dioxide. Lesser amounts of carbon disulfide, carbonyl sulfide, and carbon dioxide were also evolved. Quantitative analysis of the off-gases by TGA-GC-MS gave the results shown in Table 1. From these data a material balance of 0.9:2.0:1.2 for C:H:S in the polymer can be calculated. This ratio is in good agreement with the theoretical value of 1:2:1 anticipated for PMS, and indicates that the accountability is sufficient to propose the mechanisms in Figs. 11-14.

Under thermal conditions the only volatile components evolved were thioformaldehyde and trithiane (presumably from trimerization of thioformaldehyde), thus indicating that PMS degrades via depolymerization or unzipping following homolytic cleavage of a C—S bond.

Kinetic Expressions for Individual Reactions Occurring in Polymer Degradation

It is frequently of interest to determine the kinetics of individual reactions occurring during the degradation of polymer systems. TGA-MS can be used to accumulate this type of data, and polystyrene provides an interesting example. In the early stages of degradation, styrene monomer and benzaldehyde are evolved simultaneously. The individual rates of evolution can be obtained by plotting the molecular ion peak intensities (m/e) 104 and 106 as a function of temperature as shown in Fig. 15. From these data the activation energies for the formation of styrene and benzaldehyde under oxidative conditions can be calculated from the Arrhenius equation as 32.4 and 29.5 kcal/mole, respectively. It is interesting to note the dependence of the degradation rate constant with temperature as illustrated in the curves shown in Fig. 15.

REFERENCES

[1]   F. Zitomer, Anal. Chem., 40, 1091 (1968).
[2]   F. Zitomer, DuPont Thermogram, 6, (1969).

[3] A. H. DiEdwardo and F. Zitomer, Fourteenth Annual Conference on Mass Spectrometry and Applied Topics, Dallas, Texas, 1969.

[4] W. F. Haddon, A. H. DiEdwardo, and F. Zitomer, Fourteenth Annual Conference on Mass Spectrometry and Applied Topics, Dallas, Texas, 1969.

[5] F. Zitomer, R. E. Sullivan, and A. H. DiEdwardo, Eastern Analytical Symposium, New York, 1971.

[6] A. H. DiEdwardo and F. Zitomer, The Society of Plastics Engineers Flammability and Safety Regional Technical Conference, Newark, New Jersey, 1971.

[7] A. H. DiEdwardo, F. Zitomer, D. E. Stuetz, and B. P. Barnes, Proceedings from Conference on Fundamentals of Flammability and Combustion of Materials, Polymer Conference Series, College of Engineering, Univ. Utah, Salt Lake City, Utah, 1972.

Cross-Linking of Methyl Silicone Rubbers.
Part II. Analysis of Extractables from Samples
Cross-Linked under Various Conditions

PHILIP M. JAMES, EDWARD M. BARRALL II, BARBARA DAWSON, and J. A. LOGAN

IBM Research Laboratory
San Jose, California 95114

ABSTRACT

During the cross-linking under themal conditions of two pre-polymers, a variety of compounds may be formed in addition to the desired cross-linked network, i.e., branched chain, large ring, and linear chain of twice pre-polymer molecular weight. In order to evaluate these processes with as little chemical complication as possible, a series of two component methyl silicone pre-polymers (Sylgard 184, 186, and 188) have been cross-linked with vinyl silicone pre-polymer at various ratios of reactive groups and concentrations in silicone oil. These rubbers were evaluated by thermomechanical analysis (TMA) and then swollen with n-hexane. The cross-link density was evaluated from TMA and hexane swelling. The extracts were studied by gel permeation chromatography. As the "catalyst" concentration (active hydrogen containing pre-polymer) was increased from 5 to 60%, the percent and average molecular weight of the hexane-soluble fraction was found to decrease for the 186 and 188 samples. No higher molecular weight fraction was extractable from the

135

184 sample. For all samples the modulus and cross-link
density increased until 10 to 30% catalyst was present.
Above this concentration these parameters dropped off
gradually. Evidence is offered to support the view that
polymerization in dilute solution promotes cross-linking.
The molecular weight changes noted for the extractables
from the 186 and 188 sample suggests that branched chain
formation predominates as a side reaction product. Little
original pre-polymer is extractable in an unreacted state.
Differential scanning calorimetry was employed to follow
the thermodynamics of the reaction.

INTRODUCTION

In Part I of this study an experimental two-component silicone
rubber was cross-linked at various component ratios and in the
presence of compatable diluting agents [1]. These two-component
poly(methylsiloxanes) condense to yield cross-links but no volatile
or nonvolatile second product. This greatly simplifies calorimetric
and mechanical studies. Given some understanding of the chemical
principals of the reaction, the two-component methylsiloxanes
should be model systems for the study of the thermodynamics of
cross-linking by modern techniques of differential thermal analysis
(DTA), differential scanning calorimetry (DSC), and thermomechanical analysis (TMA).

This study extends the previous work with an experimental
siloxane resin system to three commercially available systems:
Sylgard 184, 186 and 188. The name and numerical designations
are registered trade names of the Dow Corning Co. Each resin
system designates a "resin" and "catalyst" pair which are actually
both polymeric methylsiloxanes. One component contains a reactive
hydrogen every 50 or 100 units in place of a methyl, and the other
component contains a pendant vinyl group. The basic reaction is as
follows:

$$\left(\begin{array}{c}CH_3 \\ | \\ -Si-O- \\ | \\ CH_3\end{array}\right)_x \quad \begin{array}{c}CH_3 \\ | \\ -Si-O- \\ | \\ C \\ H \diagup \diagdown CH_2\end{array} -O \left(\begin{array}{c}C_3 \\ | \\ -Si-O- \\ | \\ CH_3\end{array}\right)_y$$

About Every 50 to 100 Groups

$$\left(\begin{array}{c}CH_3 \\ | \\ -Si-O- \\ | \\ CH_3\end{array}\right)_z \quad \begin{array}{c}H \\ | \\ -Si-O- \\ | \\ CH_3\end{array} -O \left(\begin{array}{c}CH_3 \\ | \\ -Si-O- \\ | \\ CH_3\end{array}\right)_q$$

Combined in the Presence of Heat and a Soluble Platinum Catalyst

$$\left(\begin{array}{c}CH_3\\ \mid\\ -Si-O-\\ \mid\\ CH_3\end{array}\right)_x \begin{array}{c}CH_3\\ \mid\\ -Si-O-\\ \mid\\ CH_2\\ \mid\end{array} \left(\begin{array}{c}CH_3\\ \mid\\ -Si-O-\\ \mid\\ CH_3\end{array}\right)_y$$

$$\left(\begin{array}{c}CH_3\\ \mid\\ -Si-O-\\ \mid\\ CH_3\end{array}\right)_z \begin{array}{c}CH_2\\ \mid\\ -Si-O-\\ \mid\\ CH_3\end{array} \left(\begin{array}{c}CH_3\\ \mid\\ -Si-O-\\ \mid\\ CH_3\end{array}\right)_q$$

The work with these three systems is concerned with the following:

1. Molecular weight distribution of reactants and extractable
products.
2. Effect of alterations in proportions of reactive species on the
heat of reaction, cross-link density, and elastic modulus (Young's) of
the rubber. The cross-link density and modulus were determined by
both TMA and hexane swelling. The equations for the necessary calculations have been given in detail in the previous study [1].

EXPERIMENTAL

Gel Permeation Chromatography (GPC)

Due to the relative insolubility of both the pre-polymers and the
extracted polymer after cure, GPC of these materials is difficult.
Chloroform is a reasonably good solvent. Solutions of 0.2% solids in
chloroform were ultrafiltered and automatically injected into a Waters
Associates 200 GPC in 2 ml volumes. The chromatograph was equipped
with five columns covering the porosity range 1×10^6 Å (3 columns), $1 \times$
10^4 Å, and 1×10^3 Å. The mobile phase was chloroform. Data was
acquired and digitized with a time share IBM 1800 computer system.
Both the auto-inject and on-line data acquisition system have been
described elsewhere [2]. After the acquisition the data was reduced
off-line with a modified Pickett-Cantow program. The chromatograph
was calibrated with the usual narrow molecular weight polystyrene
fractions.

The molecular weights expressed in this paper are styrene equivalent
molecular weights. If it is assumed that the hydrodynamic radii of

polystyrene and poly(methylsiloxane) are equivalent, the molecular
weight averages calculated will be uniformly 28.8% too high. This is
based upon the ratio of monomer molecular weights, 74 for dimethyl-
siloxane and 104 for styrene.

This error has not been corrected in the tables and figures since
relative molecular weights are of interest. Truly accurate molecular
weights would require calibration with narrow fractions of poly(methylsiloxane). In addition the extracts from cross-linked samples are
almost surely branched. This complication will introduce large errors
in the absolute value of the GPC-determined molecular weight. It is
well known that branched molecules have an effective hydrodynamic
radius much smaller than a linear molecule of equivalent weight. These
limitations must be held in mind when considering GPC data on the
systems of interest in this paper. However, by comparing molecular
weight trends and various systematic variables it is possible to extract much useful data from the GPC study.

Extraction Studies

Samples of silicone rubber were formed in aluminum foil cups.
Each sample contained sufficient finished product to produce two
samples for swelling and one sample for TMA. The samples for swelling were weighed dry and placed in 50 ml of spectral grade hexane and
covered. This hexane contained no detectable solid residue in 500 ml
of solvent. The samples were swollen for 2 days, removed from the
hexane, blotted twice, and weighed. The balance atmosphere was
saturated with hexane vapors. Care was exercised to lose the minimum
volume of hexane in the transfer. The temperature during swelling
averaged $25 \pm 1°C$. After the samples had swollen to constant volume,
the hexane was carefully decanted into a weighted planchette, the
swelling flask washed with more hexane, and the contents of the
planchette evaporated to dryness at $35°C$ in a vacuum oven. The
planchette was reweighed and the percent extractable was determined.
As stated above, this residue was taken up in chloroform for the GPC
study. It is necessary to exercise some care with certain silicone
formulations to exclude small chips of rubber which break off during
swelling. These cause an insignificant error in the swelling weights
but a serious error in apparent residue weights. The swelling data
were converted into cross-link density and elastic modulus using the
following equations:

$$v_2 = (1/\rho_r)/[(w/\rho_h) + (1/\rho_r)] \tag{1}$$

where v_2 is the volume fraction of the polymer in the swollen state;
ρ_r is the density of the rubber = 0.9926 g/cc, ρ_h is the density of
hexane, and w is the grams of hexane sorbed per gram of rubber.

$$\nu_e = -[\,2.303 \log(1 - v_2) + v_2 + \chi_1 v_2^2\,]/V_1\,(v_2^{1/3} - v_2/2) \qquad (2)$$

where ν_e is the moles of effective network chain per cubic centimeter for tetrafunctional sites, χ_1 is the solvent/polymer interaction parameter = 0.48(4), and V_1 is the molar volume of hexane = 130.6 cc/mole.

$$\nu_e = E/3RT \qquad (3)$$

where E is the elastic modulus (Youngs), R is the gas content = 8.314 $\times 10^7$ ergs/mole °K, and T is the temperature in °K.

NMR and Titration Studies

NMR spectra were obtained on a Varian T-60 spectrometer using tetramethylsilane as a lock and carbontetrachloride as the solvent. The uncross-linked samples were brominated by the method of Uhrig and Levin [3]. The samples were titrated with a standard bromine solution in glacial acetic acid. The titration solvent was chloroform. The end point was detected by the color of excess bromine. Difficulty was experienced using more conventional potentiometric methods due to the insolubility of the silicones in the titration solvent. The results from the direct bromination are shown in Table 1. Obviously, the process is neither simple nor direct. The products of bromination were examined by NMR and IR.

Thermomechanical Analysis

The spherical indentation measurements were made with a Du Pont TMA using a probe of radius 0.1416 cm. Loads from 10 mg to 5 g were employed, the range being determined by the softness of the sample. The elastic modulus was calculated as described previously [1] with the equation

$$F = \frac{16}{9} \, Er^{1/2}p^{3/2} \qquad (4)$$

where F is the force, r is the probe radius, and p is the penetration. In practice the load in grams was graphed versus the TMA penetration to the 3/2 power. Load was the y-axis. Thus $F/p^{3/2}$ = slope $\times$ 980. The 980 translates the force into dynes/cm^2. Given the modulus, the cross-link density can be calculated from Eq. (3).

TABLE 1. Bromination of Siloxane Resin Components

Component	Average number of monomer units between pendent vinyls
Sylgard 184 resin	4,780
Sylgard 186 resin	15,000
Sylgard 188 resin	17,700
Sylgard 184 catalyst	2.8
Sylgard 186 catalyst	15.1
Sylgard 188 catalyst	5.1

DSC Studies

The DSC technique has been described in detail previously [1].
Briefly, a Perkin-Elmer DSC 1-B was employed for all measurements
of heat reaction temperature and activation energy. The heating rate
was 20°C/min. Sample weights were chosen so that a substantially
constant deflection was obtained at 4 mcal/sec sensitivity. The samples
were heated in open aluminum planchettes. The samples were made
from the same identical mixture as employed for the TMA and swelling
samples. The DSC data was acquired on an IBM System 7 computer
using a program developed at IBM. The kinetic data were calculated
by the techniques of Barrett [4] and Rogers and Morris [5]. These
are Methods I and II in the tables.

RESULTS

Chemical Composition Studies

As this study progressed it became obvious that the chemistry of
these commercial silicones was somewhat more complex than had
initially been supposed. This is most clearly demonstrated in the
results of NMR and bromination experiments.

The bromination data shown in Table 1 for the resin systems is
about as would be expected, i.e., the right order of magnitude. The
vinyl content is so low as to be at the limits of detection. However, the
statistics of the experiment were such that it can be stated with
confidence that the 184 resin has three times the amount of unsaturation
as the 186 and 188 systems. The NMR spectra of resins 184, 186, and
188 showed the expected patterns of methyl protons in the presence of

silicone. The vinyl protons are far too low in concentration to be
detected by conventional NMR. No other resonances were noted.

The bromination results on the catalyst systems cannot be accounted for on the basis of the expected chemistry. The results in
Table 1 are several thousand times too high. Other reactions must
have taken place. The NMR curves on the catalyst systems were
also unusual but similar to one another. The curve for catalyst 184
is shown in Fig. 1. The expected methyl protons are present in
addition to resonances at 4.65 and 5.85 ppm. The small peaks at 0.7
and 1.2 ppm are spinning side bands. The superimposed high sensitivity scan at 5.85 ppm may show pendant vinyl groups. These were
not expected in the catalyst.

The 4.65 ppm resonance was examined by studying the partially
brominated products (35 and 75%). The NMR spectra are shown in
Figs. 2 and 3. The 4.65 ppm resonance is very sensitive to bromination. Although not shown in this example, the 5.85 ppm resonance
disappeared completely early in bromination. A concurrent IR study
showed that for the 35% brominated materials, the silyl groups were
reduced by ~33% over the unbrominated material. This would account
for the high values given by bromination as shown in Table 1. At
present we are unable to account chemically for this process. Obviously,

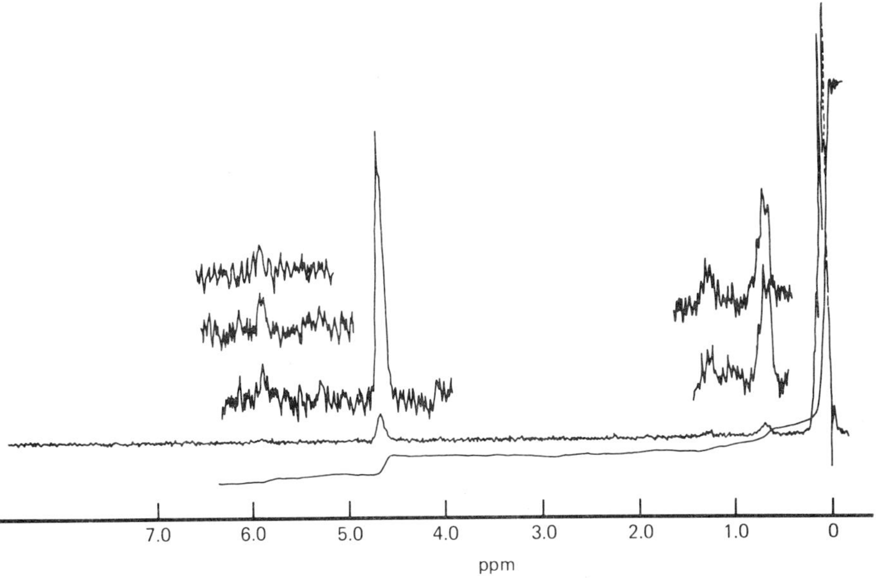

FIG. 1. NMR spectrum of Sylgard 184 catalyst showing
brominatable proton resonances.

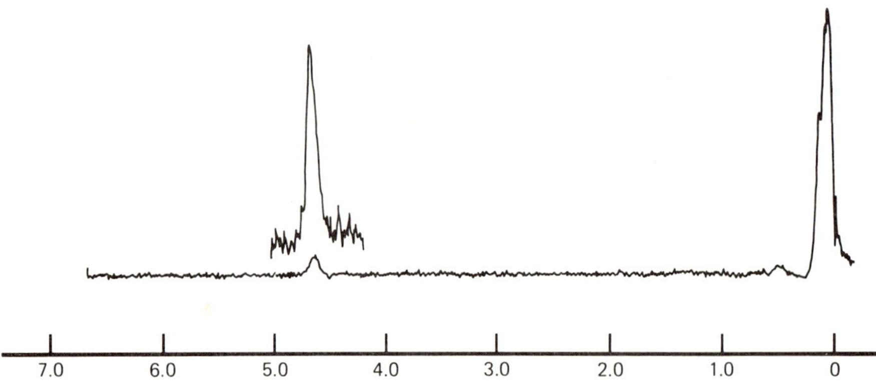

FIG. 2. NMR of 35% brominated Sylgard 184 catalyst.

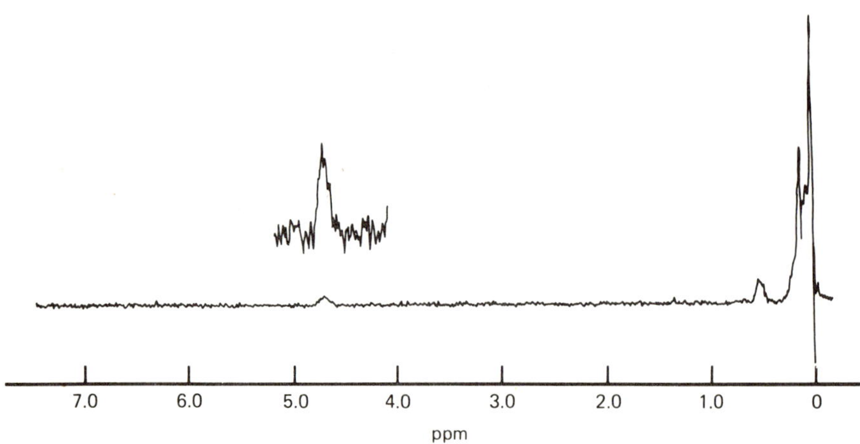

FIG. 3. NMR of 75% brominated Sylgard 184 catalyst.

bromination cannot be used to determine the labile hydrogen content in this material.

Gel Permeation Studies

The uncross-linked resins and catalysts were chromatographed and all produced bimodal molecular weight distributions. These are given in Table 2, and examples of the 184 resin and catalyst are shown in

TABLE 2. Gel Chromatographic Results on Uncross-linked Resin and Catalyst

	Mode 1			Mode 2	
Sample	Peak $\overline{M}_W$[a]	U	% Sample[b]	Peak $\overline{M}_W$[a]	Sample[b]
184 Resin	49000	1.4	85	2000	15
184 Catalyst	47000	1.3	20	1000	80
186 Resin	31000	1.6	97	700	3
186 Catalyst	55000	1.2	20	2000	80
188 Resin	57000	1.1	92	3000	8
188 Catalyst	69000	1.2	20	2000	80

[a]Weight-average molecular weight.

[b]Calculated on the basis of an equal refractive index increment for polymer and oil.

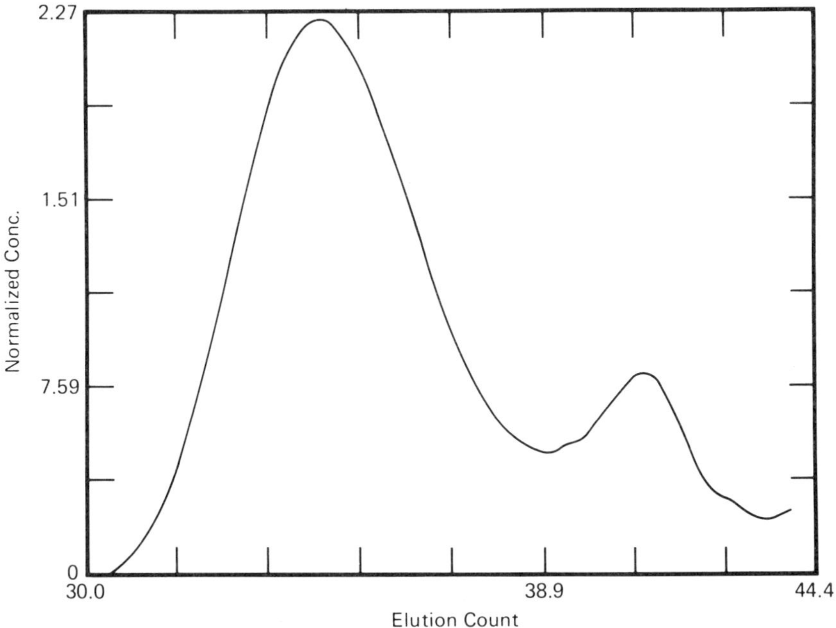

FIG. 4. Gel permeation chromatogram of Sylgard 184 resin in chloroform.

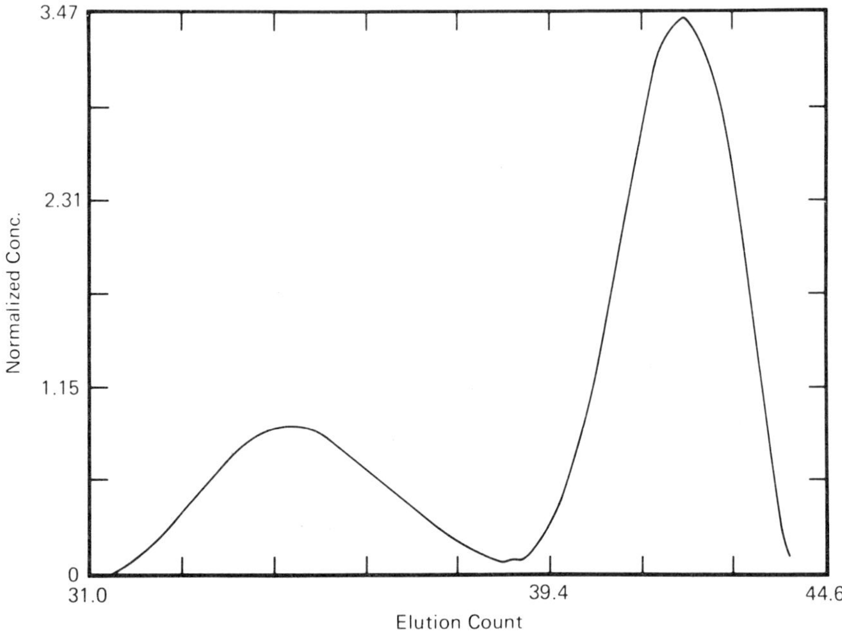

FIG. 5. Gel permeation chromatogram of Sylgard 184 curing agent in chloroform.

Figs. 4 and 5. The first peak is the high molecular weight pre-polymer, and the second peak is a low molecular weight diluting oil. NMR and IR spectral examination of the second peak indicates that it is identical to a relatively inert methyl silicone oil. The oil is probably added to adjust the viscosity of the combined resin and catalyst for various molding operations. From earlier work it may be supposed that the oil will alter the rate and completeness of cross-linking [1]. Except for the low molecular weight of the 186 resin component, there is nothing unusual about the molecular weights of the starting materials. The 186 resin is the most concentrated pre-polymer. The number of functional groups must be subject to some variation. As stated previously, we were unable to make a satisfactory direct measurement of the vinyl or active hydrogen content by several methods (NMR, IR, bromine titration). This is probably due to the low concentration of the active groups. The manufacturer indicates that one reactive group occurs per 100 monomer units. In that case the 186 catalyst should have twice as many groups as the 186 resin. The 184 and 188 resins and catalysts should be about equal. The 188 pair should present a more viscous mixture prior to

TABLE 3. Characteristics of Hexane Extractable Materials after Polymerization

Catalyst content	5	10	15	20	30	40	50	60
			Sylgard 184 System					
Total solubles, %	7.9	3.6	3.9	4.8	7.1	9.4	12.6	16.9
1st mode $\overline{M}_w$ [a]	-	-	-	-	-	-	-	-
Concentration, %	-	-	-	-	-	-	-	-
2nd mode $\overline{M}_w$ [a]	1948	1687	1904	1881	1896	1391	3076	3274
Concentration	100	100	100	100	100	100	100	100
			Sylgard 186 System					
Total solubles, %	35.5	6.4	5.5	6.1	6.8	7.8	8.4	9.4
1st mode $\overline{M}_w$ [a]	56000	47000	27000	29000	31000	33000	33000	33000
Concentration, %	90	15	5	6	8	10	15	20
2nd mode $\overline{M}_w$ [a]	1400	1400	1400	1400	1400	1200	1100	1200
Concentration %	10	85	95	95	92	90	85	80
			Sylgard 188 System					
Total solubles, %	41.0	15.0	5.6	3.8	3.8	4.7	6.5	8.5
1st mode $\overline{M}_w$ [a]	37000	23000	33000	37000	33000	33000	33000	33000
Concentration, %	75	20	5	5	3	3	5	5
2nd mode $\overline{M}_w$ [a]	1600	1400	1500	1400	1400	1400	1500	1500
Concentration, %	25	80	95	95	97	97	95	95

[a]Weight-average molecular weights.

polymerization than the 184 mixture, due to the higher molecular weight of the diluting oils.

The resin/catalyst mixtures of various concentrations show a wide variation in the percent and the molecular weight of the extractables (see Table 3). With the exception of the 184 system, all extracts are bimodal. The 184 system from the lowest catalyst concentration studied (5%) shows no high molecular weight extract. This indicates that a completely cross-linked network is formed from the lowest concentrations of catalyst. Even without a high molecular weight portion, the total percent extractables follows the same general trend as that of the other two systems as a function of catalyst concentration. There is a minimum in the extract percent curve near the stoichiometric point between resin and catalyst as indicated by the modulus and calorimeter studies and by the manufacturer. This minimum is shown graphically in Fig. 6. At higher concentrations there is some indication that a reaction does occur to produce low molecular weight material in the 184 system. The molecular weight sharply increases above 40% catalyst to a number significantly greater than the original low molecular weight fraction present in either resin of catalyst (see Tables 2 and 3).

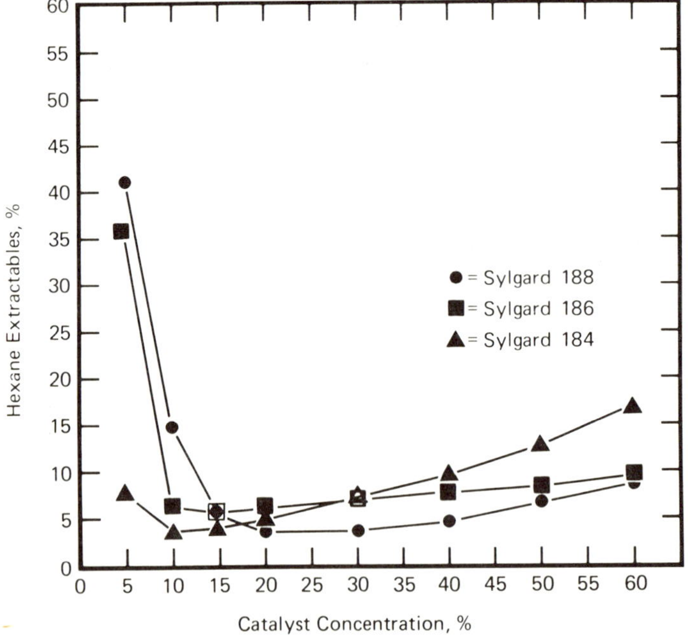

FIG. 6. Variation of extractable materials from Sylgard samples as a function of catalyst concentration.

The high molecular weight fraction predominates in the extractable portion of the 186 and 188 systems at 5% catalyst. This material is probably not related to excess resin since the molecular weight is much too high (56,000 in extract vs 31,000 in the original for the 186 resin). The extract is probably a highly branched structure with a core of excess resin molecules. The actual molecular weight could be as much as 50% higher due to branching effects in the GPC.

The percentage of high molecular weight material in the extracts drops off rapidly as the catalyst concentration is increased to the stoichiometric point. Both the measured molecular weight of the extract and the absolute amount of extract follow the same trend. This could be due to both an absolute decrease in high molecular weight material not held in a cross-linked network and to entrapment of molecules (physical) in a tightly cross-linked cage. The latter case is reinforced by the decrease in the low molecular weight fraction. In the previous study pure silicone oils were blended with the resin catalyst mixture prior to cross-linking [1]. It was not possible to extract more than 60 to 70% of these oils in hexane near the stoichiometric point.

Above the stoichiometric point the high molecular weigh portion of 186 and 188 extractables assumes a relatively constant value, ~33,000, as the resin content is increased. The low molecular weight oil predominates, but the molecular weight continues to increase. Since we have no absolute measure of branching, it is impossible to say if this apparently constant high molecular weight is absolutely constant or an artifact of increased branching.

Cross-link Density and Elastic Modulus Study

The data derived from the TMA and hexane swelling experiments are shown in Table 4. The same general trends noted previously with other systems apply to these data [1]. However, some notable exceptions are evident. The agreement between the same parameter (modulus or cross-link density) calculated from TMA and swelling is reasonably good for the 186 system. However, the cross-link density calculated by swelling is in serious error for the 184 and 188 systems.

On the basis of the physical appearance of the 20% catalyst sample of Sylgard 188, it is safe to say that the cross-link density is not 111.8×10^{-5} moles/cc. This density should be expected to produce a hard glass, which the sample surely is not. The error probably resides in the choice of the wrong value for χ_1 in Eq. (2).

The value of 0.48 had proven satisfactory for other silicone rubber systems [6] as well as the 186 system. Since the factors affecting the solvent/polymer interaction parameter are exceedingly complex, it is useless to speculate at this stage upon the reasons for the

TABLE 4. Modulus and Cross-link Density Correlation for a Series of Sylgard Rubbers

% Catalyst	Elastic modulus[a] ($dynes/cm^2$)		Cross-link density[b] (moles/cc)		Sample
	TMA	Swelling	TMA	Swelling	
5	5.4	1.1	7.4	15.4	Sylgard 184
10	11.8	66.3	16.2	90.7	
15	9.5	55.9	12.9	75.2	
20	5.4	38.0	7.4	52.1	
30	6.4	31.6	8.8	43.2	
40	8.7	34.7	11.7	47.5	
50	3.5	36.9	4.8	50.6	
60	3.0	30.1	4.2	41.2	
5	0.6	2.8	0.8	3.8	Sylgard 186
10	7.5	8.4	10.0	11.5	
15	9.5	7.7	13.0	10.5	
20	6.1	6.4	8.3	8.8	
30	5.1	4.8	7.0	6.5	
40	5.2	3.8	7.1	5.2	
50	5.5	2.8	7.5	3.9	
60	6.9	2.2	9.5	3.0	
5	0.5	1.3	0.7	1.8	Sylgard 188
10	2.3	6.5	3.1	8.9	
15	8.5	25.0	12.0	34.2	
20	8.8	81.8	12.2	111.8	
30	19.0	77.2	26.0	105.7	
40	9.6	53.2	13.0	72.8	
50	8.6	44.1	11.7	60.4	
60	6.3	35.7	8.7	48.9	

[a]Modulus $\times 10^{-6}$, i.e., $5.4 = 5.4 \times 10^6$ $dynes/cm^2$.
[b]Cross-link density $\times 10^5$, i.e., $7.4 = 7.4 \times 10^{-5}$ moles of effective cross-link/cc.

problems noted in the 184 and 188 systems. However, since this parameter is multiplicative, the values have a relative usefulness within a given system.

The variation of elastic modulus (TMA) and apparent cross-link density (swelling) with catalyst concentration are shown in Figs. 7, 8, and 9. The maximum modulus and maximum cross-link density do not always occur at the same catalyst concentration (see Figs. 8 and 9). This could be due to the contribution to the TMA modulus of entangled and/or branched chains which are not detected by hexane swelling. The secondary maximum on System 184 at 40% catalyst may be due to the same effect. System 188 appears to be very sensitive. There is some reflection of this in the data on the extracts (see Table 3).

Calorimetric Data

The heat of reaction (exothermal) is very dependent on catalyst concentration in the presence of excess resin. However, from Fig. 10 it is obvious that other reactions than that shown in the Introduction can occur. The maximum heat of reaction occurs at concentrations of

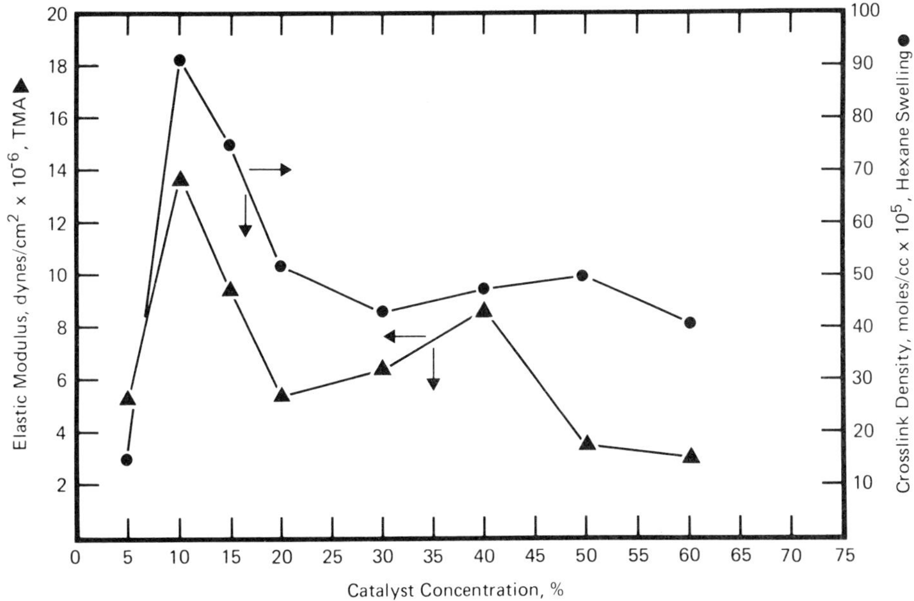

FIG. 7. Variation of elastic modulus and cross-link density with catalyst concentration Sylgard 184.

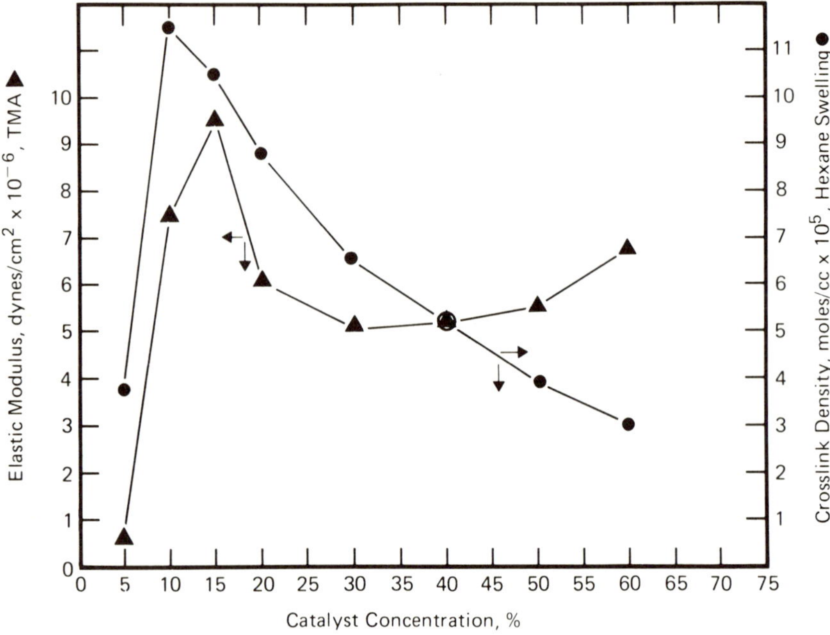

FIG. 8. Variation of elastic modulus and cross-link density with catalyst concentration Sylgard 186.

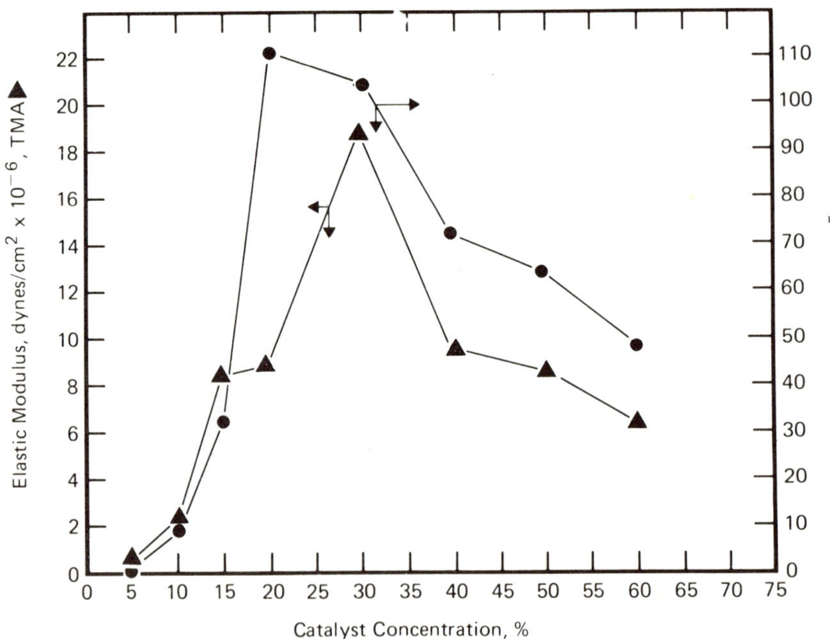

FIG. 9. Duration of elastic modulus and cross-link density with catalyst concentration Sylgard 188.

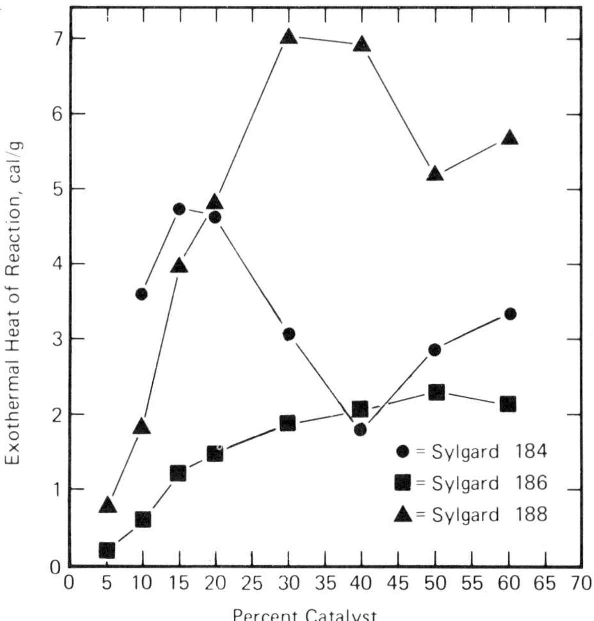

FIG. 10. Effect of catalyst concentration on the heat of reaction of three Sylgard systems.

catalyst above that required to induce the highest modulus of elasticity (compare Figs. 7 to 10). Indeed, reasonable correlation exists between maximum cross-linking and maximum heat of reactions except for the 186 system. The 186 system shows a slowly increasing reaction heat above 15% catalyst. If all of the systems had the same functionality and cross-linked to the same extent, then these results should form a system of congruent curves. This is not the case. Secondary reactions, probably involving the methyl groups, are important.

The energy of activation also shows a systematic variation with catalyst concentration (see Table 5). The activation energy was calculated by two methods as discussed previously. For this series of samples agreement between the two methods, one using partial areas and the other using rates calculated by the curve height, is reasonably good. This was not the case in the previous study [1]. A portion of the error observed in that study must be ascribed to errors encountered in the determination of the base line. The authors are unable to account for the shifts of the exothermal maximum temperature with catalyst concentration. This could be due to an alteration in the path of the

TABLE 5. Thermodynamic Data on Three Sylgard Systems

System	Catalyst concentration (%)	Heat of reaction (cal/g)	Exothermal peak temp (°C)	Energy of activation	
				Method I [4] (kcal/mole)	Method II [5] (kcal/mole)
Sylgard 184	10	3.64	97.5	46.5	42.8
	15	4.75	97.4	51.4	47.6
	20	4.65	92.9	84.3	79.4
	30	3.09	88.6	72.8	67.7
	40	1.76	91.1	69.9	63.4
	50	2.87	96.6	55.9	51.1
	60	3.34	99.3	42.8	39.6
Sylgard 186	5	0.20	103.3	40.9	31.3
	10	0.55	96.0	56.1	51.0
	15	1.24	111.4	47.8	42.9
	20	1.45	109.3	42.7	36.8
	30	1.86	110.4	43.8	38.2
	40	2.06	112.7	39.7	35.6
	50	2.30	117.0	36.5	33.0
	60	2.15	118.5	31.8	28.2
Sylgard 188	5	0.72	107.0	34.0	28.7
	10	1.80	113.8	37.9	31.6
	15	3.93	114.1	41.0	35.3
	20	4.82	113.4	36.6	32.5
	30	7.03	109.3	49.6	45.8
	40	6.95	112.0	53.1	49.1
	50	5.17	114.5	52.1	46.5
	60	5.71	115.0	45.6	42.0

TABLE 6. Gelation of a 10% Catalyst Sylgard 184
System at 25°C

Time from mixing (hr)	Heat of reaction (cal/g)
0	4.75
3	4.25
5	3.0

reaction as a function of catalyst concentration. NMR did indicate
vinyl groups on the catalyst. It is very probable that intrachain cross-
linking is significantly different kinetically from interchain processes.

The reaction producing a cross-linked system does occur at a
finite and appreciable rate at 25°C. The results of allowing a 10% catalyst mixture of the 184 system to age show this clearly (see Table 6).
The formation of an open gel at low temperature significantly alters
the structure of the network formed at elevated cure temperatures on
the aged samples (see Table 7). The modulus and cross-link density
of samples held at 25°C for 5 hr and then cured by programmed heating
generally show a decrease in most cases. However, the weight extracted
by hexane and molecular weight distribution of the extracts show only
small effects. It is proposed that gelation at low temperatures results
in an open network. In this network many cross-linkable groups remain
unreacted upon heating due to restriction by location. This behavior
would place serious restrictions on the pot life of these systems for
applications where the elastic properties are critical.

CONCLUSIONS

The commercial Sylgard systems, which are the subject of this study,
are significantly different from the silicone rubber reported in an
earlier study. In every aspect of this investigation there is good indication that secondary reactions are significant. From an applied viewpoint,
significant alterations in the thermal and mechanical properties have been
observed when either the ratio of components is varied 10 to 20% or the
duration of the reaction is changed. The initial assumption that these
systems could furnish models for the convenient study of cross-linking
processes must be seriously revised. The results of the GPC study may
have some application to the general theory of network formation. However, this picture is complicated by branching and secondary reactions.

This particular approach to the analysis of a reactive polymer system,

TABLE 7. Effect of 5 hr 25°C Treatment on Final Properties of Three Sylgard Systems

System	Elastic Modulus TMA $\times 10^{-6}$ cynes/cm^2	Cross-link density swelling $\times 10^5$ (moles/cc)	% Extractable	Molecular weight of extract	
				Mode I	Mode II
184 fresh	11.8	90.7	3.6	–	1700
aged	6.9	80.7	4.0	–	1800
186 fresh	7.5	11.5	6.4	46000	1400
aged	3.8	9.3	8.7	41000	1400
188 fresh	7.5	8.9	15.0	23000	1400
aged	3.0	9.5	13.7	33000	1600

DSC, GPC, and NMR, are valuable. In an applied case the data obtained by these means should afford improved quality control and reliability.

REFERENCES

[1] E. M. Barrall II, M. A. Flandera, and J. A. Logan, Thermochim. Acta, 5, 415 (1973).
[2] A. Gregges, B. Dowden, E. Barrall, and T. Horikawa, Separ. Sci., 5, 731 (1970).
[3] K. Uhrig and H. Levin, Anal. Chem., 13, 90 (1941).
[4] K. E. J. Barrett, J. Appl. Polym. Sci., 11, 1617 (1967).
[5] R. N. Rogers and E. D. Morris, Anal. Chem., 38, 412 (1966).
[6] A. M. Bueche, J. Polym. Sci., 15, 97 (1955).

A Kinetic Investigation of Thermal Shrinkage of Aromatic Polymers by Thermomechanical Analysis

HIROTARO KAMBE, TEIJI KATO, and MASAKATSU KOCHI

Institute of Space & Aeronautical Science
University of Tokyo
Komaba, Meguro-ku, Tokyo, Japan

ABSTRACT

The kinetics of thermal shrinkage of the stretched poly-pyromellitimide (PI) films, Du Pont Kapton H, were investigated by thermomechanical analysis (TMA) at a constant rate of heating. The two-stage model of extended polymers has been applied to analyze the TMA data. The activation energy of the contraction reaction could be obtained from TMA curves at various heating rates. The one-step shrinkage of the 12.5% stretched PI film gives an activation energy of 10 kcal/mole for the first shrinking. From the 30% stretched sample, the second activation of contraction is 25 kcal/mole, and from 40% sample the third one is 33 kcal/mole. These three contraction reactions are observed successively for the TMA curves for highly stretched samples, and correspond to the various kinds of molecular motion of this special rigid polymer structure, which are also observed in the dynamic mechanical and dielectric properties of the same polymer.

157

Polypyromellitimide (PI) is a typical thermally stable polymer due to the rigid aromatic and heterocyclic ring structures in its backbone chain. This polymer is able to be cold-drawn in the glassy state, and it shrinks markedly on heating. The thermal shrinkage was investigated thermoanalytically by thermomechanical analysis (TMA) at a uniform heating rate. The activation energies of thermal shrinkage in three stages are estimated by this method.

EXPERIMENTAL

The sample is the commercial polypyromellitimide film, Du Pont Kapton H, with the basic structure

The specimen was stretched up to 50% under tension with a constant rate of stretching at room temperature. The TMA apparatus, as shown in Fig. 1, was a modified linear expansion apparatus of Rigaku Denki Co. with a balance-type loading and detection units. The TMA specimen was cut from an extended film sample to a thin strip. The change in length of the specimen was recorded continuously in vacuum at the various heating rates under a tensile load of 10 g. The creep of the film was negligible for this load.

THERMAL SHRINKAGE

The TMA curves at $10°C/min$, obtained for cold-drawn samples with different degrees of stretching, are shown in Fig. 2. The ultimate degree of shrinkage increases and the temperature range in which the sample shrinks is broadened with the degree of stretching. Figure 3 shows the temperature derivative curves of the same data. The thermal shrinkage of the 12.5% stretched sample constitutes a one-step reaction with a maximum rate at $90°C$. In the 20 and 25% samples, another reaction was observed at $200°C$, and more stretched samples show a distinct contraction at $350°C$. Therefore, in highly stretched samples, three contraction reactions occur successively.

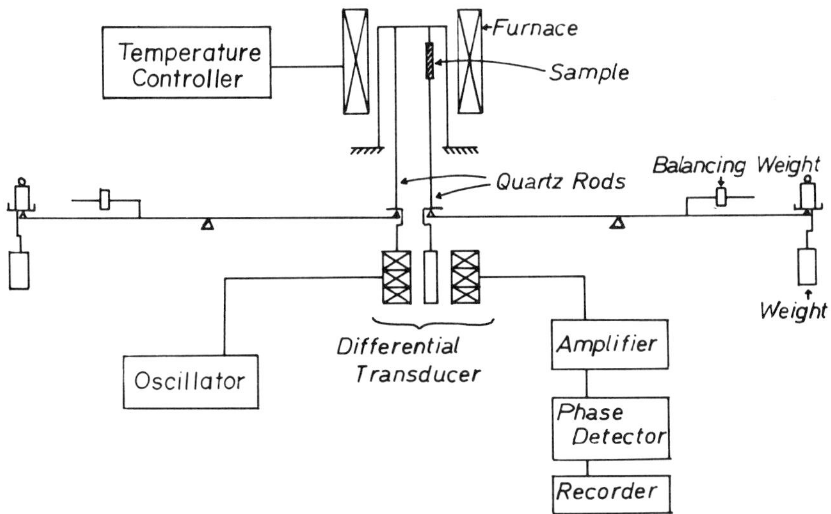

FIG. 1. Schematic diagram of TMA apparatus.

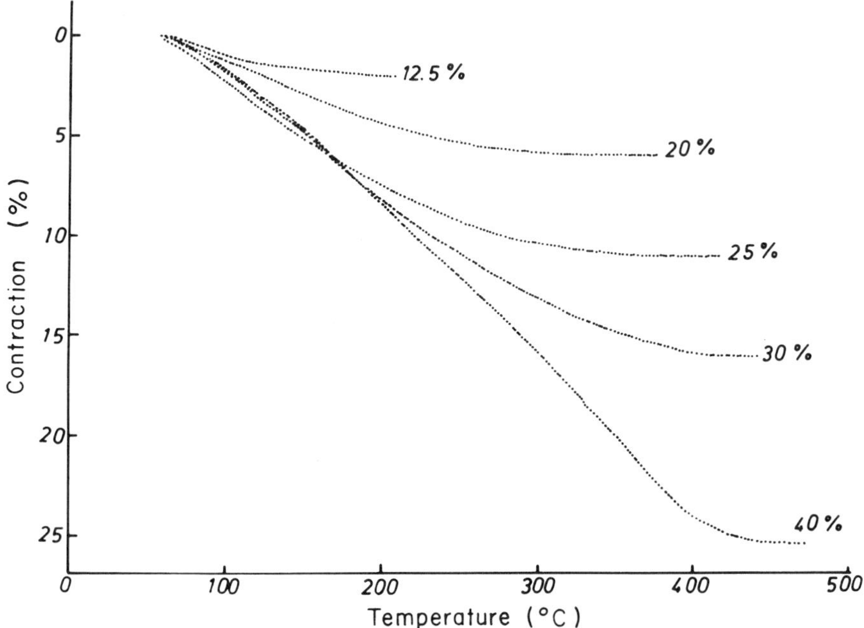

FIG. 2. Thermal contraction of extended PI films with various degrees of stretching.

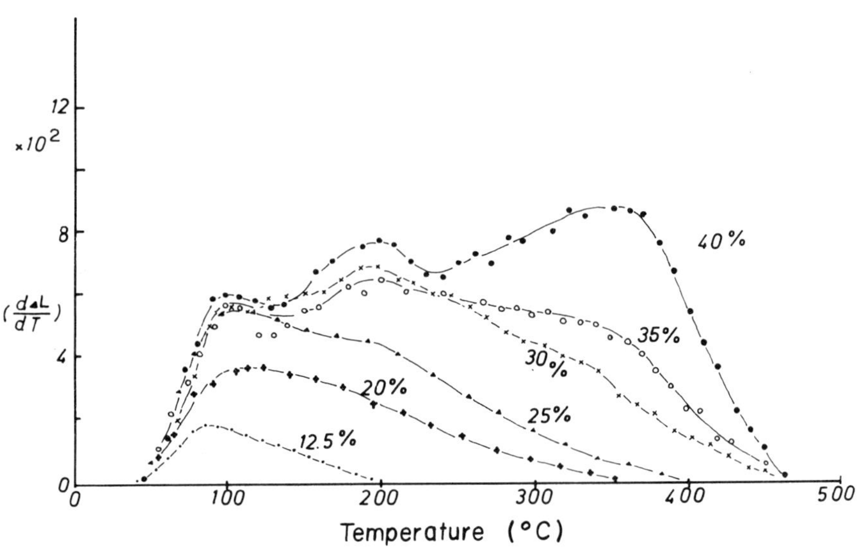

FIG. 3. Temperature derivative curves of thermal contraction.

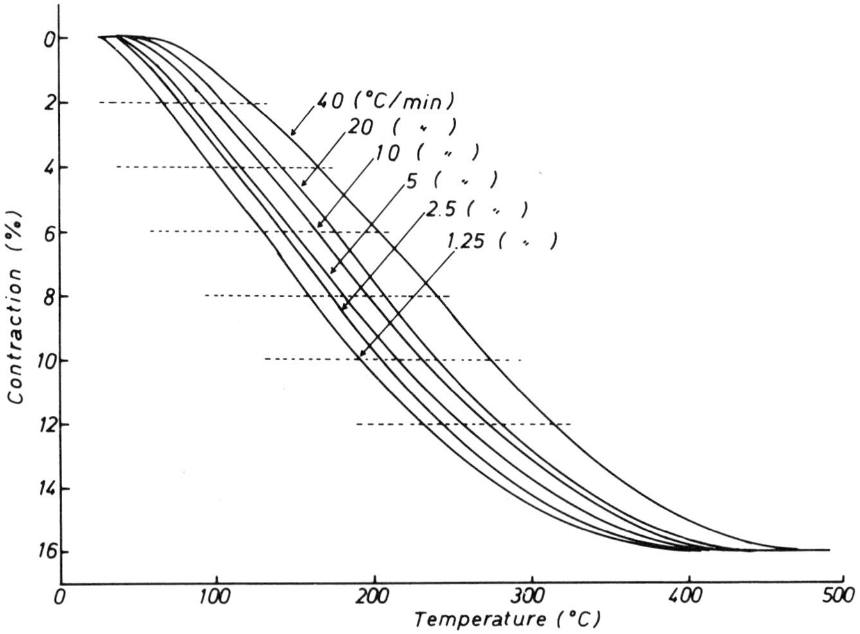

FIG. 4. TMA curves of PI films stretched by 30%.

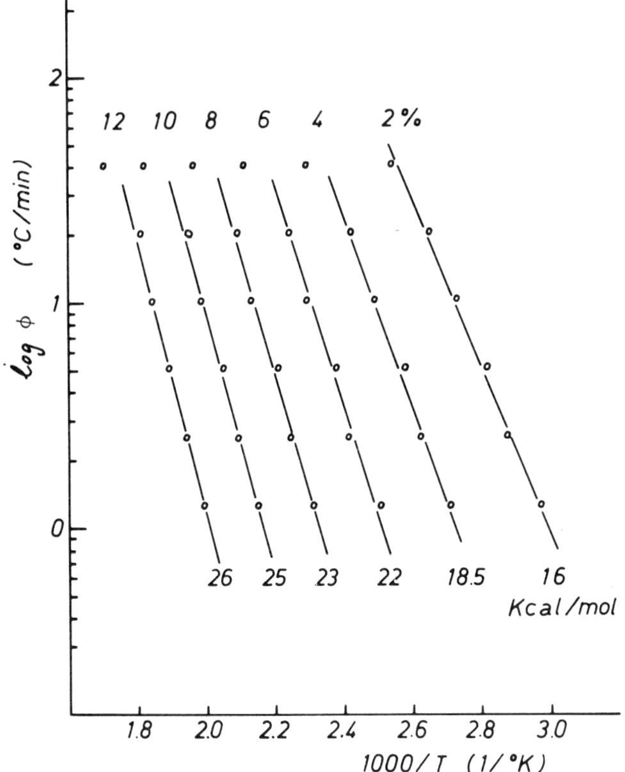

FIG. 5. Kinetic plots for 30% stretched sample.

THEORETICAL ANALYSIS OF TMA CURVE

A two-state model of the extended molecule is assumed for the analysis of TMA curves. This analysis is to be published elsewhere [1], but a brief summary is given here. We consider the lower α and the upper β states to correspond to the normal and extended states of segments. At time t in thermal shrinkage, the number of segment fractions changed from β to α is taken as x. The rate of shrinking is expressed by a simplified form with the Arrhenius equation:

$$\frac{dx}{dt} = A \exp\left(-\frac{\Delta E}{RT}\right)g(x) \tag{1}$$

By integrating, with a uniform heating rate ϕ of TMA,

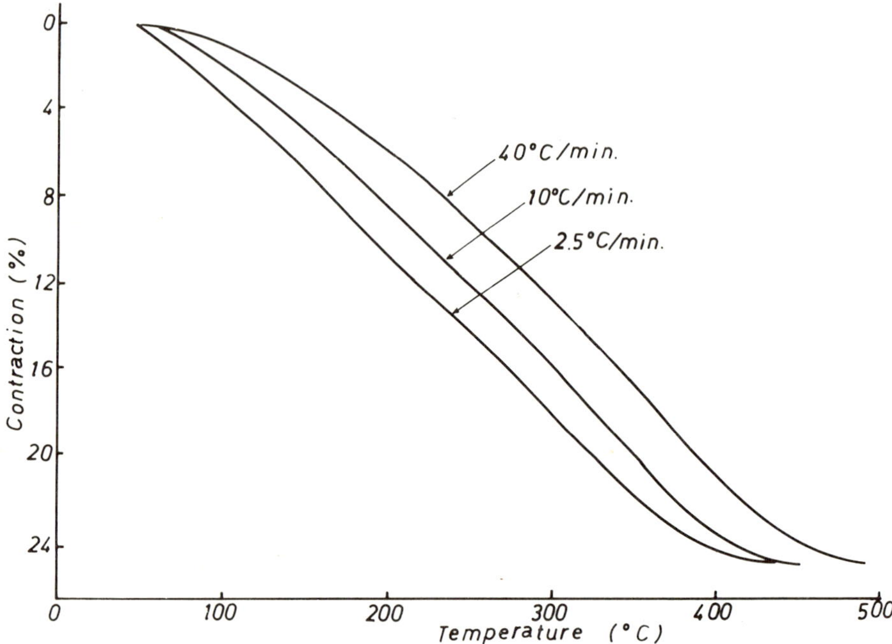

FIG. 6. TMA curves of PI films stretched by 40%.

$$G(x) = \int_0^x \frac{dx}{g(x)} = A\theta \tag{2}$$

where

$$\theta = \frac{1}{\phi} \int_0^x \exp\left(-\frac{\Delta E}{RT}\right) dT \tag{3}$$

Let y be $\Delta E/RT$, and integrating by parts

$$\theta = \frac{\Delta E}{\phi R} \left[\frac{\exp(-y)}{y} + Ei(-y)\right] = \frac{\Delta E}{\phi R} P(y) \tag{4}$$

where

$$Ei(-y) = -\int_y^\infty \frac{\exp(-y)}{y} \, dy \tag{5}$$

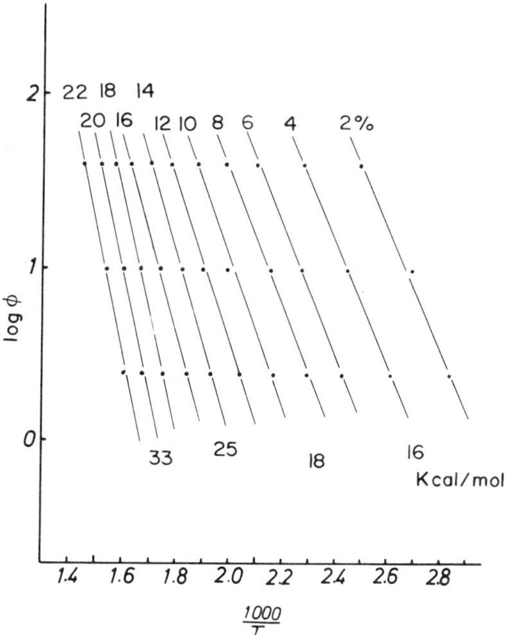

FIG. 7. Kinetic plots for 40% stretched sample.

Equation (4) is very familiar in the kinetic analysis of thermogravimetry. By using Doyle's approximation [2], we obtain

$$\log \phi_1 + 0.4567 \frac{\Delta E}{RT_1} = \log \phi_2 + 0.4567 \frac{\Delta E}{RT_2} = \ldots$$

where T_1, T_2, etc. are obtained at a degree of contraction from TMA curves for various heating rates ϕ_1, ϕ_2, etc. Therefore, if $\log \phi$ is plotted against $1/T$, we obtain the activation energy of the contraction reaction by the slope of the straight line.

RESULTS AND DISCUSSION

The kinetic analysis of the TMA curve for one-step shrinkage of 12.5% stretched film gives an activation energy of 10 kcal/mole for this reaction. Figure 4 shows the TMA curves at various heating rates for a 30% stretched sample. From them, we obtain straight lines for the log ϕ vs $1/T$ curves at various contraction levels, as seen in Fig. 5, which give an activation energy of 25 kcal/mole for the second contraction reaction. The values from lower contraction levels are somewhat lower because of the effect of the first reaction. From Figs. 6 and 7 for a 40% stretched sample, the activation energy is estimated as 33 kcal/mole for the third contraction. These three contraction reactions correspond to the loss maxima in the dynamic mechanical and dielectric measurements [3], and they suggest the existence of different mechanisms of contraction of the rigid polyimide chains.

REFERENCES

[1] H. Kambe and T. Kato, Appl. Polym. Symp., 20, 365 (1973).
[2] C. D. Doyle, J. Appl. Polym. Sci., 6, 639 (1962).
[3] H. Kambe and T. Kato, Presented at the IUPAC Symposium on Macromolecular Chemistry, Aberdeen, Scotland, September, 1973.

The Effect of Volume Reorganization of Amorphous Poly(ethylene Terephthalate) on Thermal Properties

C. C. YAU,* W. K. WALSH, and D. M. CATES

Department of Textile Chemistry
North Carolina State University
Raleigh, North Carolina 27607

ABSTRACT

When poly(ethylene terephthalate) was quenched from above T_g and then heated, it exhibited a step increase in thickness in the glass transition region at every rate tested. When the polymer was cooled more slowly than it was heated, a higher T_g and a slightly larger step increase in thickness were observed as the cooling rate was reduced. These experimental results appear to be adequately interpreted on the basis of the normal structural changes that occur in a glass as its thermal history is varied. Two observations, however, were not easily included in this view. First, the polymer, on cooling from above T_g, exhibited an abnormally high expansion coefficient over much of the range of temperature in which it exists as a fluid. Second, the polymer exhibited a step increase in thickness when it was heated at the same rate at which it had previously been cooled.

*Present address: Emery Industries, Inc., Cincinnati, Ohio 45232

165

INTRODUCTION

At any stage of its existence a polymer carries the imprint of its past. During fabrication into its various forms of fiber, film, etc., the segments of the polymer molecules are frequently "set" through the use of temperature, stress, and solvent in configurations and orientations, and confined to volume elements that are unable to adjust to equilibrium when the polymer is cooled more rapidly to the glassy state than the segments can relax. The polymer thus has a potential for change which may or may not occur at imperceptible rates depending upon the temperature. This capacity for change may be demonstrated by annealing the polymer at temperatures below the glass temperature (T_g). The polymer increases in density and, upon reheating, exhibits endothermal peaks near T_g by differential scanning calorimetry (DSC) and step increases in volume by dynamic volume dilatometry [1, 2]. These results have been interpreted as possibly indicating some first-order character in the glass transition [1-3].

Other authors [4-6] conclude that the absorption of thermal energy can be accounted for on the basis of the kinetics of the glass transformation, so that only the normal structural changes that accompany the approach to the equilibrium glassy state need be postulated. The excess enthalpy and volume trapped on cooling are held to be responsible for the subsequent behavior. Petrie [5] points out that the amount of energy absorbed is a function not only of the extent of enthalpy relaxation that occurs during the period of annealing and of the higher T_g which results from the relaxation, but also of the rate of testing. The energy absorbed thus depends on the temperature and time of annealing, and on the rate of heating during the thermal scan as well.

In the present study, most of the volume changes that may be made to occur with poly(ethylene terephthalate) in the vicinity of its glass transition by varying its thermal history appear to be satisfactorily interpretable on the basis of this latter hypothesis.

EXPERIMENTAL

Amorphous poly(ethylene terephthalate) (PET) was melt-extruded film, 50 mil (1.34 mm) in thickness, and was reported to have an intrinsic viscosity of 0.57 in phenol-tetrachloroethane at 30°C and a carboxyl end-group content of 35 meq/kg. Examination of the film between crossed polars revealed only slight orientation. The density of the as-received sample in a gradient column of sodium bromide solution was 1.3374 g/ml.

Thermal analysis was carried out by means of the Perkin-Elmer

Model TMS-1 Thermomechanical Analyzer (TMA) and the Perkin Elmer Model 1-B Differential Scanning Calorimeter (DSC). When the TMA was used, the reading of the thermocouple indicating sample temperature was recorded directly. The glass transition temperatures were corrected for thermal lag. The change in film thickness with temperature was detected by TMA. Volume changes with temperature were followed by the volume dilatometer, the construction and operation of which have been described in detail by Bekkedahl [7].

RESULTS AND DISCUSSION

A representation of a typical thermogram obtained on the TMA with PET film is shown in Fig. 1. Region a represents the glassy state; Region b represents the step increase in thickness that is observed once T_g is exceeded; Region c represents the thermal expansion path followed by the fluid. T_g is taken as the intersection of the tangent lines to a and b. A measure of the jump in thickness may be obtained from the vertical distance between this intersection and the intersection obtained from tangent lines to b and c.

The as-received sample film expanded along the path indicated by Fig. 1. In the glassy Region a, the polymer had the normal linear expansion coefficient of $0.77 \times 10^{-4}/°C$. The step increase in thickness (Region b) varied somewhat from one sample to another, between 1 and 3% of the original thickness. Most of this increase is believed to be due to a slight orientation put into the film at the time of fabrication. Once the as-received sample has undergone this step increase, it is taken to be completely isotropic. In this case the relationship between linear and volume expansion coefficients is given by $3\alpha_1 = \alpha_v$, where α_1 is the linear expansion coefficient and α_v is the volume expansion coefficient. When the sample was again heated in the TMA, a step increase

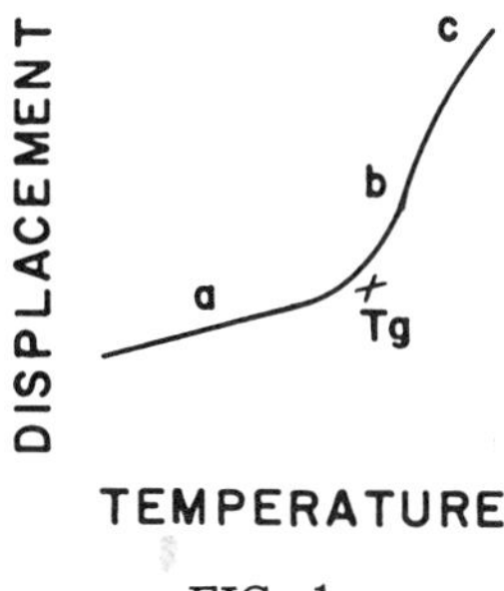

FIG. 1.

in thickness was again seen, but the increase amounted only to 0.2 to 0.4% of the sample thickness. This increase has its origin in the conditions of testing that obtain after the original orientation has been removed. In the work to be discussed below, the necessity for considering the prior history of the sample was eliminated by subjecting the sample to a heat treatment at 90°C. At higher temperatures the sample may undergo a significant degree of crystallization.

The linear expansion coefficient of the polymer in Region c (Fig. 1) was difficult to measure accurately because only a limited interval of temperature is available before the rate of crystallization becomes appreciable. A value of $4.9 \times 10^{-4}/°C$ for a heating rate of 20°C/min was estimated, which is much larger than the value of $1.7 \times 10^{-4}/°C$ that is characteristic of the isotropic fluid. It appears that under the conditions of the test, i.e., in this case a temperature rise of 20°C/min, the sample was unable to reach the equilibrium fluid state, and was therefore expanding at a much greater rate than it would if the expansion were between two states in thermal equilibrium.

Thermal Behavior of the Quenched Sample

When the sample was quenched from 90°C to below T_g and reheated in the TMA at different programming rates, the step increase of Region b was again observed at each rate.

The step increase in thickness appeared to vary somewhat with programming rate, but the procedure for determining the increase was not sufficiently precise to allow the values at the various rates to

TABLE 1. Expansion Behavior of Quenched PET

Rate of heating ($°C$/min)	T_g ($°C$)	Expansion coefficient $\times 10^4$	
		Below T_g $(°C)^{-1}$	Above T_g $(°C)^{-1}$
40	91	0.70	-
20	88	0.69	-
10	83	0.77	-
5	78	0.79	-
2.5	77	0.77	-
1.25	76	0.74	-
Volume dilatometer $(\alpha_1 = \frac{1}{3}\alpha_v)$	67	0.77	1.7

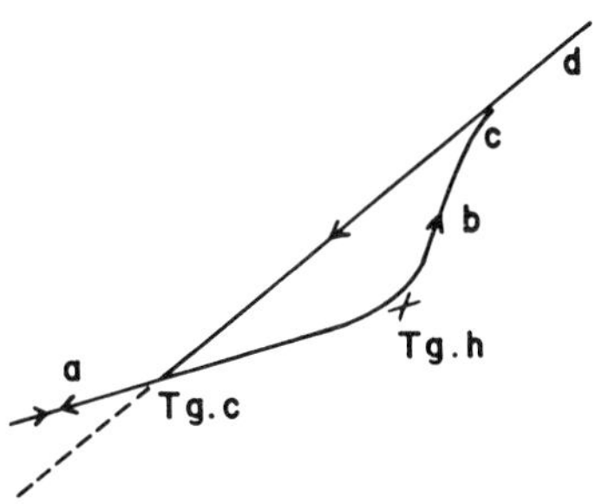

TEMPERATURE

FIG. 2.

be accurately differentiated. As indicated by Table 1, T_g increased with the rate of the test, as expected [5], and the expansion coefficient in the glassy region was approximately the same at all rates. The experimental results may be interpreted with the help of the hypothetical thermogram of Fig. 2. On cooling, the fluid is transformed to the glass at $T_{g.c}$. Subsequently, when the glass is heated at a particular rate, it is converted to the fluid at $T_{g.h}$. The higher the rate of temperature rise, the greater $T_{g.h}$. T_g may be defined as the temperature at which the response time of the molecular segments equals the time interval of the test. An increase in rate of temperature rise reduces the interval of the test. It therefore becomes necessary to reach a higher temperature, i.e., $T_{g.h}$, with each increase in rate of temperature rise in order to reduce the response time of the segments sufficiently to achieve equality with the time interval of the test.

Thus in Region b the polymer, now existing as a fluid, expands, as rapidly as segmental relaxation times permit, to equilibrium thickness. The rate of expansion drops off somewhat in Region c and approaches asymtotically the equilibrium rate of the fluid.

One would expect that if the sample were heated at an extremely slow rate, so that the polymer could approach the equilibrium free volume at each temperature more closely than was permitted by any of the rates for which a step increase in thickness was observed, the sample would not exhibit the step increase and would have a much lower T_g than was observed at the higher rates. This procedure was followed by volume dilatometry on the as-received sample. In the TMA the sample exhibited the behavior pictured by Fig. 1. In the volume dilatometer, with the temperature being adjusted for each point in the transition region at a rate not exceeding $0.1°C/min$, the volume exhibited no discontinuity, T_g had the low value of $67°C$, and the volume expansion coefficients in the glassy and fluid regions had the normal values of $2.3 \times 10^{-4}/°C$ and $5.3 \times 10^{-4}/°C$, respectively.

It should be pointed out that the step change in thickness of the as-received sample near T_g, which was earlier attributed to a slight orientation, was not observed as a volume change in the volume dilatometer because, as has been shown previously [8], a dimensional increase in the direction perpendicular to the plane of orientation is compensated by decreases parallel to the plane.

Expansion Behavior for Equal Cooling and Heating Rates

It may be expected [4] that if the polymer fluid were first cooled at a particular rate, the polymer would lie at the same distance from equilibrium at each temperature during the descending and ascending parts of the cycle. In this case the polymer should exhibit neither endotherms nor step increases in thickness when first cooled and then heated at the same rates. This expectation, however, does not appear to be borne out experimentally. Thus for temperature regimes ranging from 40 to $1.25°C/min$ (cooling rate = heating rate), the PET sample exhibited during the heating part of the cycle not only step increases in thickness on the TMA (Fig. 1) but also endotherms on the DSC (Fig. 3). This result suggests that the polymer undergoes hysteresis between cooling and heating although the cooling and heating rates are equal, i.e., T_g on cooling ($T_{g.c}$) is less than T_g on heating ($T_{g.h}$) for each rate tested. Such an effect, if it were real, would mean that the mobility of the molecular segments at equal temperatures within the range from $T_{g.c}$ to $T_{g.h}$ is different during the descending and ascending legs of the cycle. Although the authors are unaware of any theoretical basis for such an effect, the possibility was tested by subjecting the polymer fluid to consecutive stages of cooling and heating. A representation of the type of thermogram obtained is presented in Fig. 4, which is interpreted in the following way. The polymer exists as the fluid above T_g. During the initial cooling, Region e is observed, in which the linear expansion coefficient appears to be much greater than the accepted value of $1.7 \times 10^{-4}/°C$. The explanation for this apparent step change, which is qualitatively similar to that observed on heating, is not known. Region d represents the interval of temperature over which the expansion coefficient corresponds approximately to the accepted value of the fluid. This interval was rather small when the test rate was $5°C/min$ or higher, but became larger as the test rate was lowered.

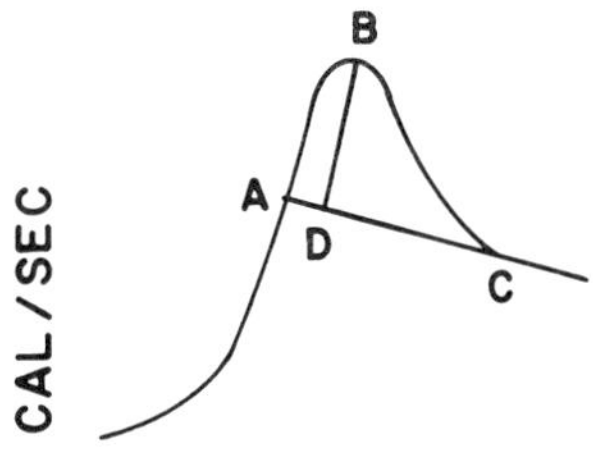

FIG. 3.

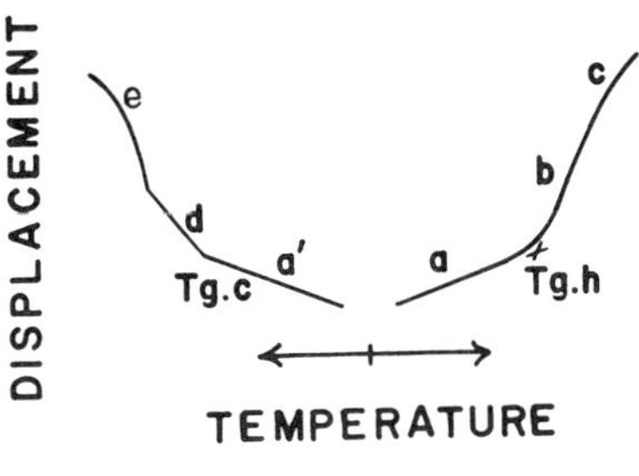

FIG. 4.

The experimental results are given in Table 2. The difference between $T_{g.c}$ and $T_{g.h}$ is appreciable at each of the rates cited. However, both $T_{g.c}$ and $T_{g.h}$ are expected to vary with rate: $T_{g.h}$ changes with rate whereas $T_{g.c}$ does not. It is conceivable, given the original premise, that the mobility of the polymer during cooling is sufficient at each of the rates employed to allow a close approach to equilibrium so that $T_{g.c}$ does not change significantly with rate in this range. However, the rate of change of $T_{g.h}$ with decreasing heating rate is so small as to make it uncertain that $T_{g.h}$ would extrapolate to the $T_{g.c}$ value of 69°C at a very low heating rate. Some doubt must exist, therefore, as to whether $T_{g.c}$ and $T_{g.h}$ may be compared when they are determined by the procedures that have been outlined.

TABLE 2. Expansion Behavior of PET Subjected to Equal Cooling
and Heating Rates

| Rate | $\alpha_1 \times 10^4 \; (^\circ C)^{-1}$ | | | | |
($^\circ$C/min)	Region d	Region a'	Region a	$T_{g.c}$ ($^\circ$C)	$T_{g.h}$ ($^\circ$C)
5	1.9	0.80	0.75	69	78
2.5	1.6	0.74	0.74	69	77
1.25	1.8	0.73	0.73	69	76

Expansion Behavior for Cooling Rate Less Than Heating Rate

The amount of free volume entrapped in a polymer decreases as the
cooling rate decreases (4, 5). Therefore, the T_g observed at a given
heating rate should increase for a polymer that is first subjected to
successively lower cooling rates from above T_g. According to Fig. 2,
the glassy Region a would intersect the equilibrium fluid line d at
successively lower temperatures ($T_{g.c}$) as the cooling rate is reduced.
The glass temperature observed during heating at the given rate
($T_{g.h}$) would then be raised by an amount sufficient to correspond with
the reduction in free volume produced by the particular, prior cooling
rate. Thus slow cooling produces glass with a higher density, which,
for the given heating rate, would exhibit a higher T_g. One sees from
Fig. 2 that this combination of higher density and higher T_g would, for
the given heating rate, tend to make the step change in thickness somewhat greater.

A study was carried out in which prior cooling rates of 1.25 and
20°C/min were used with a heating rate of 20°C/min to above T_g. $T_{g.h}$
was about 2°C higher for the case in which the prior cooling rate was
1.25°C/min. Also the step increase in Region b (Fig. 1) was somewhat
greater with the lower cooling rate; 0.39% compared to 0.27% based on
the original thickness.

This same experiment was repeated on the DSC, using a 15-mg
sample of PET. The results are given in Table 3.

The thermogram of Fig. 3 indicates the procedure used for the
measurements: BD represents peak height, and ABC represents the
area used in calculating the energy of the transition. The densities
were obtained by the gradient method using sodium bromide solution

TABLE 3. Effect of Cooling Rate on Thermal Properties of PET; Heating Rate of 20°C/min

Prior cooling rate (°C/min)	Density at 30°C (g/ml)	T_g (°C)	Peak height (mm)	Energy (mcal)
1.25	1.3372	75.3	35	4.65
20	1.3364	73.3	8	0.75

on replicate samples immediately after they were cooled at the prescribed rate.

The greater loss of free volume by the slow-cooled sample is shown by its greater density. The size of its endotherm relative to that of the fast-cooled sample (4.65 mcal compared to 0.75 mcal) is more pronounced than the difference in step change in thickness obtained on the TMA.

It is also possible, by first annealing the polymer below T_g, to produce the same effect as by slow cooling [1, 4, 5]. The greater the time of annealing at a particular temperature (or for a given time of annealing the higher the temperature of annealing), the larger the endotherm. The time and temperature of annealing were observed to have this same effect on the step change in thickness that occurs in the glass transition region. These experimental results may be explained on the basis of the molecular relaxation that occurs as a result of the annealing [5].

SUMMARY

The polymer, when quenched from above T_g and heated, exhibited a step increase in thickness in the glass transition region at every rate tested, presumably because the relaxation rate of the segments did not allow the expansion path of the equilibrium fluid to be followed. At the slow rate of temperature adjustment that was used in the volume dilatometer, no step increase in volume was observed. When the polymer fluid was cooled and then heated at the same rate, an endotherm on the DSC and a step increase in thickness on the TMA were observed. The explanation for this behavior is not known, but it is suggested that the polymer in a dynamic test may undergo hysteresis between cooling and heating. When the polymer was cooled more slowly than it was heated, a higher T_g and a slightly larger step increase in thickness were observed as cooling rate was reduced. This result was attributed to smaller amounts of free volume being entrapped in the polymer at the lower cooling rates.

Most of the experimental results appear to be adequately interpreted on the basis of the normal structural changes that occur in a glass as its thermal history is varied. Two observations, however, were not easily included in this view. First, the polymer, on cooling from above T_g, exhibited an abnormally high expansion coefficient over much of the range of temperature in which it exists as a fluid. Second, the polymer exhibited a step increase in thickness when it was heated at the same rate at which it had previously been cooled.

ACKNOWLEDGMENT

We wish to thank Dr. C. J. Heffelfinger, E. I. du Pont de Nemours and Co., Inc., Circleville, Ohio, for samples of poly(ethylene terephthalate). Thanks are also due to Mrs. R. Kirby and Mr. G. D. Chambers for TMA and volume dilatometric measurements.

REFERENCES

[1] M. S. Ali and R. P. Sheldon, J. Appl. Polym. Sci., 14, 2619 (1970).

[2] G. W. Miller, Amer. Chem. Soc., Org. Coatings Plastics Chem. Preprints, 32(1), 146 (1972).

[3] P. V. McKinney and C. R. Foltz, J. Appl. Polym. Sci., 11, 1189 (1967).

[4] A. E. Tonelli, Macromolecules, 4, 653 (1971).

[5] S. E. B. Petrie, J. Polym. Sci., A-2, 10, 1255 (1972).

[6] R. C. Roberts and R. F. Sherliker, J. Appl. Polym. Sci., 13, 2069 (1969).

[7] N. Bekkedahl, J. Res. Nat. Bur. Stand., 42, 145 (1949).

[8] C. C. Yau, W. K. Walsh, and D. M. Cates, Polym. Preprints, 13, 1181 (1972).

Thermo-Optical Analysis of Poly(2,6-disubstituted-1,4-phenylene Oxide) Blends

A. R. SHULTZ and B. M. GENDRON

General Electric Corporate Research and Development
Schenectady, New York 12301

ABSTRACT

Mobility transitions closely related to the glass transitions
of three binary polymer-polymer blend systems have been
determined by thermo-optical analysis (TOA). TOA in
this instance consists of the automated observation of
birefringence relaxation in scratched transparent polymer
and polymer blend films during their programmed heating.
The blend systems studied were: I. polystyrene (PS) +
poly(2,6-dimethyl-1,4-phenylene oxide)(PMMPO); II.
poly(2-methyl-6-phenyl-1,4-phenylene oxide)(PMPPO) +
PMMPO; and III. PS + PMPPO. Blend Systems I and II
display a single TOA transition temperature at each blend
composition indicating homogeneity and true thermodynamic
compatibility of the polymer pairs. Blend System III displays two TOA transition temperatures at each blend
composition indicating incompatibility and resultant liquid-
liquid phase equilibration in molten mixtures of PS and
PMPPO.
The shapes of the transmitted light intensity vs temperature
curves for scratched PS + PMMPO films between 90° crossed
plane polarizer and analyzer are examined at three heating rates
of the microscope hot stage. The use of green light rather than

white light in the analysis was examined briefly in System II
blends. Glass transitions T_g(DSC) obtained by differential
scanning calorimetry are also reported.

INTRODUCTION

The use of thermo-optical analysis (TOA) to determine mobility
transitions in amorphous blends of polystyrene (PS) + poly(2,6-
dimethyl-1,4-phenylene oxide)(PMMPO)[1], poly(co-styrene/p-
chlorostyrene) + PMMPO [2], and poly(2-methyl-6-benzyl-1,4-
phenylene oxide) (PMBPO) + PMMPO [3] has been reported. TOA
as presently employed is a rather empirical, automated observation
of birefringence relaxation in stressed/strained transparent polymer
films during heating as an indicator of molecular mobility onset. The
TOA transition temperature, T_{TOA}, defined as the temperature at
which scratch-induced birefringence in the sample completely dis-
appears, is slightly higher than the glass transition temperature, T_g,
of the sample. The operational simplicity and sensitivity of the TOA
method recommend its expanded use as a detector of glass transitions.
The complexity of the factors contributing to the absolute form and
substance of TOA data makes detailed analyses of the controlling
molecular and supermolecular physical properties difficult.

The present study was undertaken to provide a further demonstra-
tion of the utility of TOA, to examine the effect of heating rate on the
TOA curves of Blend System I (PS + PMMPO), and to examine the
nature of two new blend systems: System II, poly(2-methyl-6-phenyl-
1,4-phenylene oxide)(PMPPO) + PMMPO; and System III, PS + PMPPO.

EXPERIMENTAL

Materials and Film Preparation

The polystyrene used was Lot PS4a (Pressure Chemical Co.), an
anionically-polymerized polystyrene having a nominal molecular weight
97,200 and an M_w/M_n = 1.06. The PMMPO (an experimental PPO
resin, General Electric Co. registered trademark) was prepared by
the oxidative coupling of 2,6-xylenol [4-7]. Its intrinsic viscosity at 30°
in chloroform is 0.49 dl/g, M_n = 18,500 by osmotic pressure and M_w =
37,200 by light scattering. The PMPPO was prepared by the oxidative
coupling of 2-methyl-6-phenyl phenol. Its [η] at 30° in chloroform is
0.88 dl/g.

Polymer blend films were prepared as follows. One-gram mixtures of polymer were dissolved in 10 ml toluene by stirring at 40°. The clear, homogeneous solutions were then precipitated into 200 ml methanol in a Waring blender. The fine, fibrous precipitates were collected on a sintered-glass frit filter, air-dried, and dried overnight at 80° in a vacuum oven. The blends thus obtained were then compression molded into films of about 5 to 8 mil thickness. The molding temperatures ranged from 180 to 270°C for the PS + PMMPO and PS + PMPPO blends. The PMPPO + PMMPO blends were molded at 240°C. Prior to precipitation of the toluene solutions of the PS + PMPPO mixtures, a drop of each solution was placed on a microscope slide and allowed to dry in air under an inverted Petri dish. This gave two sets of films for Blend System III, one compression molded and the other solution cast.

Measurement of Transition Temperatures

For comparison with TOA, some T_g measurements were made by differential scanning calorimetry. A Perkin-Elmer DSC Model 1B was uscd at a 20°/min heating rate. Its temperature scale was calibrated by observation of the melting points of indium, tin, and lead.

TOA measurements were performed as previously described [1]. Scratches scribed in the sample films with a steel stylus at room temperature provided stress/strain birefringent regions for monitoring during heating at constant rate in the hot stage (Mettler FP2 plus FP21) of a Zeiss standard WL polarizing microscope. The light that was transmitted through a plane polarizer, the sample, and a 90° crossed plane analyzer was picked up by a photocell. The amplified photocurrent was fed through a voltage divider and the voltage difference was plotted continuously against time (temperature) on a strip chart recorder (Leeds and Northrup, AZAR-Speedomax). The normal heating rate for TOA was 10°/min. Heating at 2 and 1°/min was also investigated for Blend System I. The temperature control unit reading was calibrated at each heating rate by observing the disappearance of light transmission during melting in the hot stage of small samples of naphthalene, adipic acid, and 2-chloroanthraquinone between the crossed polarizers. The results are presented in Table 1. In reporting temperatures in the TOA runs the control dial readings were adjusted by -2, -0.8, and -0.7° for the 10, 2, and 1°/min heating rate runs, respectively. White light from the incandescent tungsten lamp was normally used. A rather broad-band green light was obtained by insertion of a Zeiss #27316 filter for one series of runs on Blend System II. Normal TOA procedure at 10°/min heating rate involved scribing the film, making a run from 60 to 250°, removing the film on its slide, allowing it to cool to room temperature with the slide in contact with a

TABLE 1. Calibration of Hot-Stage Control Unit

Compound	Triple point ($^\circ$C)	T(apparent) - T(actual) (deg.)		
		1°/min	2°/min	10°/min
Naphthalene	80.24	0.4	0.7	2.3
Acipic acid	151.46	0.7	0.8	2.5
2-Chloroanthraquinone	208.95	0.8	1.0	2.6

stone bench top, re-scribing the film, and making a second run. In
some instances a third run was made. For runs at 2 and 1°/min the
films were heated at 10°/min to a temperature 30° below the T_{TOA}
at 10°/min heating rate, and then the control was switched to the
slower heating rate.

RESULTS

The TOA results will first be presented for each blend system
individually. A summary discussion concerning all three blend
systems will then follow.

I. Polystyrene (PS) + Poly(2,6-dimethyl-1,4-phenylene Oxide) (PMMPO)

TOA and DSC measurement results on this blend system were
presented in a previous publication [1]. The II blend films of this
series covered the total composition range in weight fraction incre-
ments of 0.100. This seemed a good series to examine with regard
to the effect of heating rate on the form of the apparent transmitted
light intensity vs temperature curve and on the observed transition
temperature.

Figure 1 represents the strip chart recording for the PS + PMMPO
film having weight fraction composition w_{PMMPO} = 0.800 being heated
at 1°/min. The six temperatures T_s, T_u, $T_{1/2}$, T_i, T_{TOA}, and T_E
indicated on the plot were tabulated, when possible, for each TOA curve
of the 11 sample films at each of the three heating rates. The accuracy
of determining the starting and end temperatures, T_s and T_E, was low
due to the low-angle approach of the curves to the initial and final

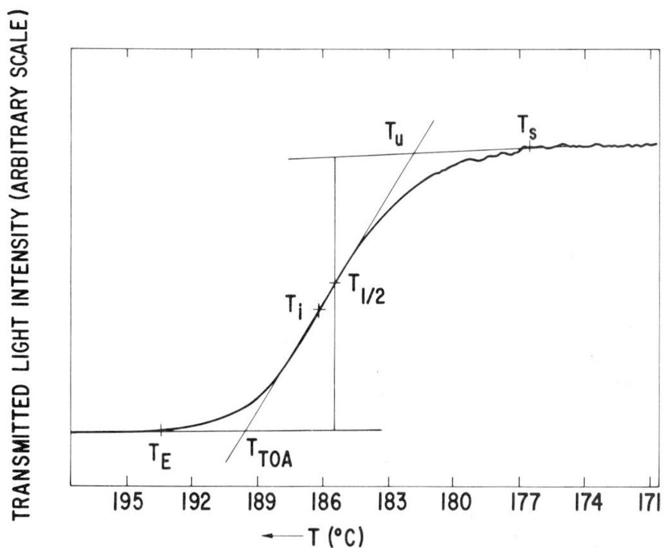

FIG. 1. TOA curve indicating possible characteristic temperatures. PS + PMMPO film having w_{PMMPO} = 0.800. Heating rate, 1°/min.

tangents. T_u, the temperature of intersection of the initial tangent with the tangent through the inflection point, suffered in accuracy due to the difficulty of locating the initial tangent. The temperatures T_i and T_{TOA}, denoting the temperature at the curve's inflection point and the temperature of intersection of the tangent through this point with the final tangent, respectively, were the most readily determined temperatures. These temperatures are plotted in Figs. 2 and 3. As one might expect, the transition temperatures observed at the slower heating rates are in general displaced to slightly lower temperatures. In the composition range $0.500 \leq w_{PMMPO} \leq 1.000$, the T_{TOA} and T_i behaviors at the three heating rates are quite orderly and consistent. No sample of the w_{PMMPO} = 0.400 film was available for the 2 and 1°/min measurements. The film having w_{PMMPO} = 0.300, for which at 10°/min the transmitted light intensity vs temperature curve exhibited a drop and rise before the final disappearance of birefringence [1], yielded no satisfactory transition temperatures at the 2 and 1°/min heating rates. The transmitted light intensities had dropped to very low values at the start of the slow heating runs (i.e., at 100°, which is 30° below the T_{TOA} at the 10°/min heating

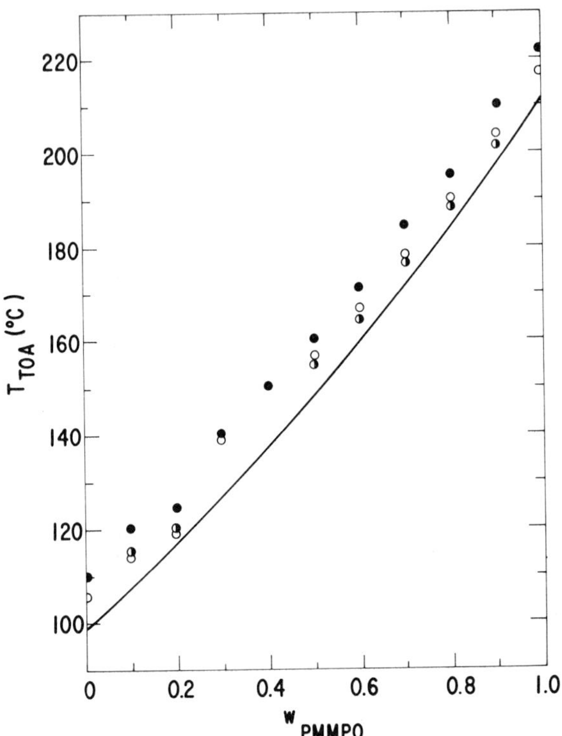

FIG. 2. TOA transition temperatures of PS + PMMPO blend films plotted against weight fraction of PMMPO. Heating rates: ($\bullet$) $10°$/min; ($\circ$) $2°$/min; ($\circ$) $1°$/min. The solid curve represents T_g(DSC) measured at $20°$/min heating rate.

rate). Although there were slight increases and declines in the transmitted light intensities, no T_i or T_{TOA} could be assigned for $w_{PMMPO} = 0.300$. It is believed that the measurement difficulty lies in the opposite signs of the stress optical and strain optical coefficients in PS plus a compensating birefringence contribution by the PMMPO. At $w_{PMMPO} = 0.200$, 0.100, and 0.000 the films display consistent T_{TOA} at the three heating rates. The T_i at $10°$/min for the films having $w_{PMMPO} = 0.100$ and 0.000 are low, reflecting the observed rather gradual decrease in transmitted light intensity with temperature prior to the final disappearance.

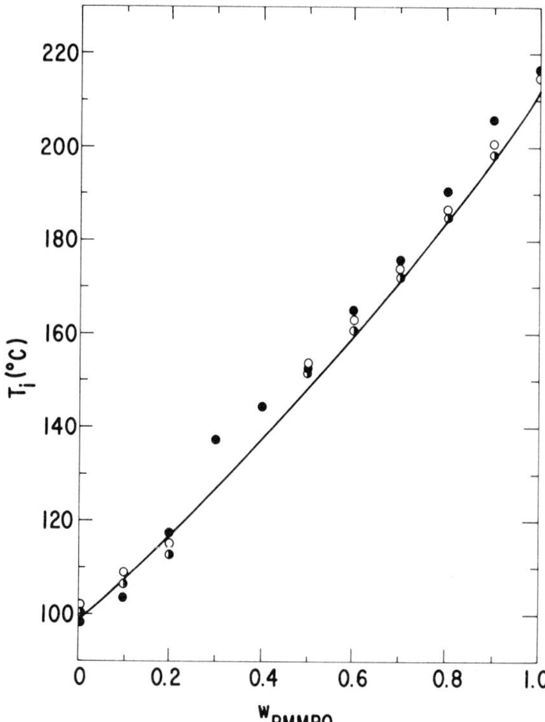

FIG. 3. Temperatures, T_i, at the inflection points of PS + PMMPO TOA curves plotted against weight fraction of PMMPO. Symbols as in Fig. 2. Solid curve represents T_g(DSC).

The curves in Figs. 2 and 3 represent the glass transition temperatures T_g(DSC) obtained by differential scanning calorimetry at a 20°/min heating rate [1]. The T_i at 2 and 1°/min approach rather closely to these T_g(DSC).

II. Poly(2-methyl-6-phenyl-1,4-phenylene Oxide) (PMPPO) + Poly(2,6-dimethyl-1,4-phenylene Oxide) (PMMPO)

Figure 4 presents the TOA plots obtained for the compression molded films of Blend System II using white light. Figure 5 presents the TOA

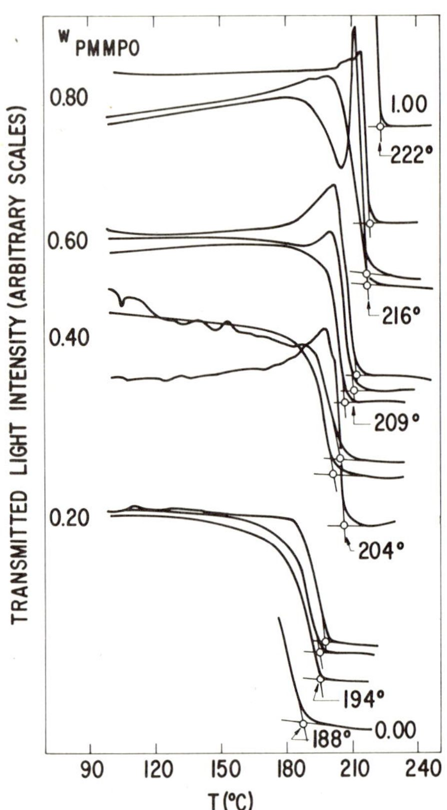

FIG. 4. TOA curves for System II (PMPPO + PMMPO) blend films. Transmitted white light; 10°/min heating.

plots (first runs only) obtained using green light. The latter were slightly smoother, but no significant differences are noted between the TOA curve shapes or the T_{TOA} values obtained with white light or green light transmission.

The measured T_{TOA} ($10°$/min) are listed in Table 2. Also listed are the T_g(DSC) observed for these blend films. Both the TOA and DSC data exhibit single transition temperatures, indicating the films to be homogeneous, single-phase blends.

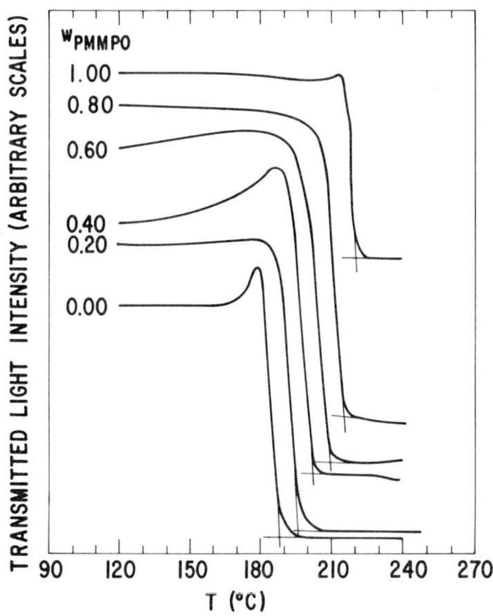

FIG. 5.   TOA curves for System II blend films. First run; green light; 10°/min heating.

TABLE 2.   Thermo-optical and Glass Transition (DSC) Temperatures in PMPPO(1) + PMMPO(2) Blends

w_2	T_{TOA}(°C)		T_g(DSC)(°C)
	White light TOA	Green light TOA	
0.000	190, 188,[a] 187[a]	188, 186	169, 170
0.200	194, 192, 196	197, 193	177, 175
0.400	206, 202, 204	202, 203	182, 181
0.600	206, 210, 210	210, 210	190, 189
0.800	215, 216, 217	215, 215	203, 203
1.000	222, 222	221, 221	211, 213

[a]Annealed at 200° for 3 min.

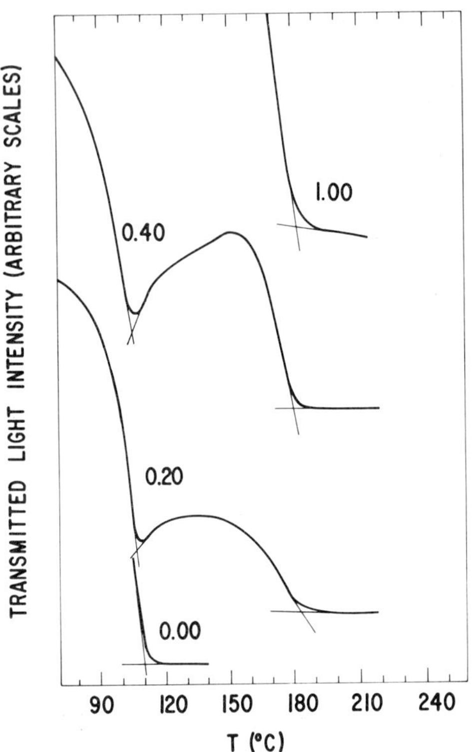

FIG. 6. TOA curves for System III (PS + PMPPO) blend films. Pure PS, pure PMPPO, and blends having weight fraction of PMPPO equal to 0.20 and 0.40. Heating rate, 10°/min.

III. Polystyrene (PS) + Poly(2-methyl-6-phenyl-1,4-phenylene Oxide) (PMPPO)

The blend films of System III all exhibited turbidities ("milky" or "cloudy") which were visually indicative of inhomogeneous, two-phase structures. TOA curves for pure PS4a, pure PMPPO, and solution cast films having weight fractions PMPPO of 0.20 and 0.40 are represented in Fig. 6. The shapes of the TOA curves for these blend films are quite similar to those observed for the incompatible blend films of PMMPO + poly(p-chlorostyrene) [2]. Two distinct T_{TOA} are noted. These correspond approximately to the T_{TOA} of the individual polymers which comprise the blends.

TABLE 3. Thermo-optical and Glass Transition (DSC) Temperatures in PS(1) + PMPPO(2) Blends

| w_2 | T_{TOA} (°C) | | | T_g (DSC) (°C) | |
	Compression molded film	Compression molded film; annealed 3 min at 200°C	Thin film cast from toluene solution	Compression molded film	Fibrous co-precipitated blend
0.00	110 116		(116) 110	91 99	
0.20	- , - 113, 185	114, 185 114, 181	100, 185 109, 181	103, - 105, 154	108, 175 103, 162
0.40	114, 193 110, 182	108, 181 107, 181	91, 180 107, 182	107, - 104, 166	94, - 106, 160
0.60	- , 178 - , 185	98, 184 103, 183	(108), 176 (121), 182	- , - - , 168	106, - 105, 168
0.80	- , 186 - , 187	(101), 183 (99), 182	- , 181 (121), 180	- , 172 - , 172	(113), 177 (99), 168
1.00	190	188 187	175 182	169 170	

The T_{TOA} and T_g (DSC) for Blend System III are given in Table 3. The T_{TOA} were measured on the compression molded films, on these same films annealed for 3 min at 200°C, and on the thin, solution cast films. The T_g (DSC) were measured on the compression molded films and on the coprecipitated blends from which these films were formed. It is somewhat difficult to detect the transitions, T_{TOA} and T_g (DSC), of the minority phase in the nonannealed, compression molded blend films. This is especially true in the detection of the lower temperature transition, and it is similar to the difficulty of detecting the T_g of a semicrystalline polymer.

DISCUSSION

The observed thermo-optical transition temperatures for the three polymer blend systems are summarized in Fig. 7. The T_{TOA} of Blend System II (PMPPO + PMMPO) are seen to increase linearly with increasing PMMPO content in the blends. The relation may be expressed as

$$T_{TOA} \; (°C) = 189 + 33w_2 \tag{1}$$

where w_2 is the weight fraction of PMMPO.

The T_g (DSC) (cf. Table 2) for System II may be fairly well represented by a least-squares fitted relation

$$T_g (DSC) \; (°C) = 170 + 43w_2 - 20w_2 (1 - w_2) \tag{2}$$

The T_{TOA} dependence on w_2 does not, therefore, exactly parallel the T_g (DSC) dependence, but ranges from 19° above T_g (DSC) at $w_2 = 0$ to 9° above T_g (DSC) at $w_2 = 1.0$. There is evidence of only one T_{TOA} and one T_g (DSC) at each blend composition. Poly(2-methyl-6-phenyl-1,4-phenylene oxide) therefore joins polystyrene [1], various styrene/p-chlorostyrene statistical copolymers [2], and poly(2-methyl-6-benzyl-1,4-phenylene oxide) [3] on the list of polymers thermodynamically compatible with poly(2,6-dimethyl-1,4-phenylene oxide).

System III (PS + PMPPO) blends exhibit incompatibility as demonstrated by two T_{TOA} and two T_g (DSC) at given blend compositions.

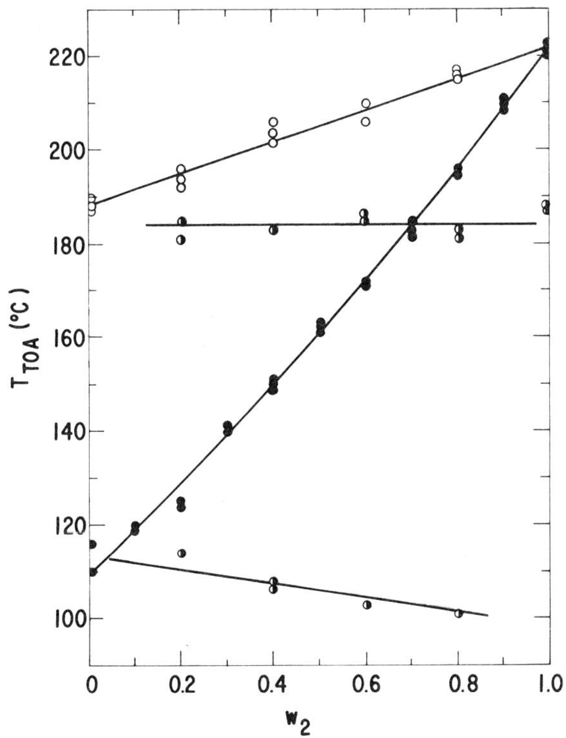

FIG. 7. Thermo-optical transition temperatures plotted against the weight fraction of Component 2. Blend systems: ($\bullet$) I, PS(1) + PMMPO(2); ($\circ$) II, PMPPO(1) + PMMPO(2); and ($\circleddash$) III, PS(1) + PMPPO(2).

There is some indication (Fig. 7) that the T_{TOA} of the phase rich in PMPPO is lowered slightly by the PS. Since the equilibrating phases in such "quasi-binary" polymer blends should contain molecules of each polymer distributed between them, there is reason to expect this [8, 9]. The effect, however, is slight in the present system. One also expects the T_{TOA} and T_g(DSC) of the phase rich in PS to be elevated by the PMPPO molecules equilibrated into it. Although the data scatter precludes a firm estimate of this effect, the T_g(DSC) data (Table 3) do indicate a possible 5 to 7° elevation of the PS glass transition by the PMPPO contained therein. Any elevation of the T_{TOA} of the phase rich in PS by PMPPO is obscured by the onset of increasing light transmission in approach to the second transition, producing an apparent shift of the transition to lower temperatures.

TOA and DSC provide complementary evidence of compatibility in polymer blends. Although TOA depends upon light as a probe, it is not limited by a need for any refractive index difference between separated phases, nor is it limited by the requirement that the domain dimensions in inhomogeneous blends be comparable to or exceed the wavelength of the light. The T_{TOA} of the matrix of a clear inhomogeneous blend can be determined without regard to the dimensions of the internal phase. The effect of orientation relaxations within and at the boundaries of very small domain internal phases upon the light transmission in the TOA measurement as presently constituted deserves theoretical and experimental exploration. As an empirical tool TOA is providing a very useful service in the examination of certain polymer blend systems. The use of birefringence observations to examine changes in submolecular and supermolecular mobility of single-component and multicomponent systems has a considerable history. The development of novel ways to apply such observations to polymer characterization remains a fertile field in polymer physicochemical research.

ACKNOWLEDGMENTS

The authors wish to acknowledge helpful discussions with Dr. G. D. Cooper concerning his early work on DSC of PMPPO + PMMPO blends. Thanks are also due to Dr. D. M. White and Mr. H. J. Klopfer who supplied the PMPPO and to Mr. R. W. Beers for assistance in film molding operations.

REFERENCES

[1] A. R. Shultz and B. M. Gendron, J. Appl. Polym. Sci., 16, 461 (1972).

[2] A. R. Shultz and B. M. Gendron, Paper Presented at the 6th Biennial Polymer Symposium, University of Michigan, Ann Arbor, Michigan, June 12-15, 1972.

[3] A. R. Shultz and B. M. Gendron, Paper Presented at the 2nd International Symposium on Polymer Characterization, Battelle Seattle Research Center, Seattle, Washington, August 21-23, 1972; J. Polym. Sci., Polym. Symp., 43, 89 (1973).

[4] A. S. Hay, U.S. Patent 3,306,875 (February 28, 1967), assigned to General Electric Co.

[5] A. S. Hay, H. S. Blanchard, G. F. Endres, and J. W. Eustance,
      J. Amer. Chem. Soc., 81, 6335 (1959).
[6] A. S. Hay, J. Polym. Sci., 58, 581 (1962).
[7] D. M. White and H. J. Klopfer, Ibid., A-1, 8, 1427 (1970).
[8] R. Koningsveld, H. A. G. Chermin, and M. Gordon, Proc. Roy.
      Soc., Ser. A, 319, 331 (1970).
[9] R. Koningsveld, Chem. Zvesti, 26, 263 (1972).

Characterization of Fibers by
Dynamic Thermoacoustical Analysis

PRONOY K. CHATTERJEE

Personal Products Company
Subsidiary of Johnson & Johnson
Milltown, New Jersey 08850

ABSTRACT

Dynamic thermoacoustical analysis has been described for
the characterization of certain physicochemical properties
of fibers. The method consists of a continuous measurement of the propagation time of constant frequency sonic
pulses transmitted through the sample which is being held
under light tension and heated under programmed temperature. The sonic pulses are generated by subjecting a
piezoelectric crystal with electrical pulses of 7 kHz
frequency. Dynamic thermoacoustical curves were obtained
with a variety of synthetic fibers and cellulose fiber. The
fibers were heated from room temperature to their respective melting points or decomposition temperature. Theoretical principles are discussed with respect to the
relationship between sonic responses and the fundamental
properties of polymers. The dynamic viscoelastic equation,
based on Maxwell-Wiechart model for polymers, has been
combined with sonic pulse propagation equation in order to
relate the sonic response with the viscoelastic parameters
of unoriented and oriented polymers. It is shown that the

sonic response under dynamic heating condition is influenced by two factors: 1) orientation of polymer molecules and 2) sonic viscoelastic function of the polymer. It has been further demonstrated that the orientation of fibers practically remained unchanged on heating under the experimental conditions examined. The dynamic thermoacoustical curves for fibers revealed only the characteristic viscoelastic properties of polymers as a function of temperature.

INTRODUCTION

Thermoanalytical techniques in the field of polymer characterization have been rapidly expanding. The instrumentation on commonly known techniques such as differential thermal analysis (DTA) and thermogravimetric analysis (TGA) have been improved to a great extent during recent years while newer methods are constantly being introduced into the field. Among the newer methods, thermomechanical analysis (TMA) and torsional braid analysis (TBA) have been found to be extremely useful in characterizing certain physico-mechanical properties of polymers. Although the overall polymer characterization capabilities have been greatly improved by these modified and newer techniques, there are still many characteristic properties of polymers which cannot be determined by the conventional thermoanalytical techniques. Therefore, the exploration of newer techniques to obtain hitherto undetermined properties of polymers is still very important.

The thermal-sonic technique, which is the subject of discussion of this paper, would determine the viscoelastic behavior of polymers under dynamic heating condition. The technique is somewhat analogous to rheovibron measurements but differs in many respects in theory and practice.

The use of sonic techniques for determining structure-property relationships in polymers was first reported by Ballou and Smith [1] in 1949 and the technique was further advanced by Charch and Moseley [2, 3].

In this work the main principle of the technique suggested by Ballou and Smith [1] has been adapted to develop a method for studying the viscoelastic properties of polymers under dynamic heating conditions. Also a theory has been developed to relate the sonic response to certain fundamental properties of polymers.

In broader terms, dynamic thermal analysis comprises techniques of measuring certain properties of material while it is being heated at constantly increasing temperature. The present technique consists of measurements of sound velocity in polymer while the polymer is being heated under a programmed rate. It therefore, falls under the general category of dynamic thermal analysis. The technique will be referred to as "dynamic Thermoacoustical analysis."

EXPERIMENTAL

Dynamic Thermoacoustical Analysis

In principle, the method consists of a continuous measurement of
the propagation time of sonic pulses of constant frequency (7 kHz)
through the sample which is being held under light tension and heated
at programmed temperature. In the present investigation the samples
were heated from ambient temperature to the respective melting
points or decomposition.

The experimental set-up is shown in Fig. 1. In the figure, A
represents the heating block of a Du Pont 900 DTA where H represents
the heating element and T_2 and T_3 represent the reference and sample
wells, respectively. Two additional holes (T_1 and T_4) were drilled all

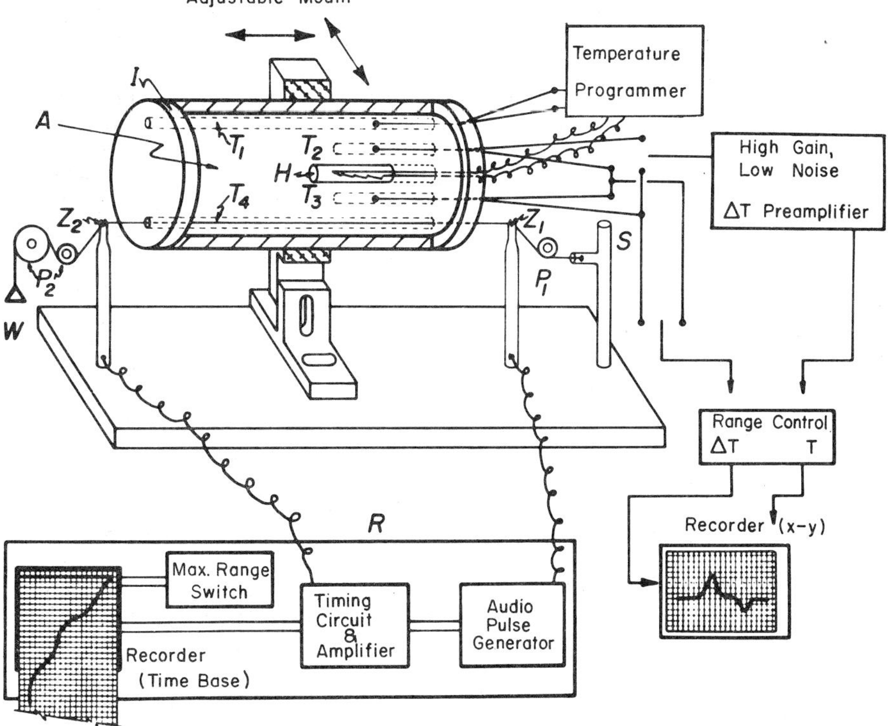

FIG. 1. A schematic representation of dynamic thermoacoustical
system.

the way from one end to the other through the heating block. The
heating block was thoroughly insulated by putting asbestos caps on
both ends and covering the rest with asbestos tape. Melting point
tubes, open on both ends, were inserted into the holes T_1 and T_4.
The heating block was mounted horizontally and thermocouples were
inserted in T_1, T_2, and T_3. These thermocouples were connected to
different terminals of the Du Pont DTA cell as shown in the Fig. 1.
The fiber sample was tied at one end to a clamp S, passed under a
pulley P_1 and on a notched ceramic piezoelectric crystal trans-
ducer Z_1, through the hole T_4, and supported by another identical
piezoelectric crystal Z_2 and a set of pulleys P_2 and finally terminated at a suspended weight of 5 g. The piezoelectric crystals were
connected to an electrical pulse generating and recording device, R.
The distance between Z_1 and Z_2 was 3.6 cm, which was kept constant
throughout the experiment. The piezoelectric transducers and the
pulse generating and recording devices were the parts of an instrument known as Pulse Propagation Meter, Model PPM-5R, manufactured by H. M. Morgan and Co., Cambridge, Massachusetts. The
maximum range switch of the instrument was modified to obtain a
time response as high as 2000 μsec.

As soon as the pulse propagation meter is activated, the electrical
pulses of 7 kHz frequency are transmitted to the piezoelectrical
crystal Z_1. The crystal lattice of Z_1 then begins to vibrate at the
same frequency as the electrical pulses, and thus the electrical pulses
are converted to sound pulses. Piezoelectricity is a means of con-
verting electrical energy into mechanical energy and vice versa. The
sound pulses are transmitted through the fiber sample to the crystal
Z_2, which reconverts the sound pulses back to electrical pulses. As
soon as this reconversion takes place, an internal timing circuit in
the instrument is closed and the recorder instantaneously records the
time required for pulses to propagate from Z_1 to Z_2. Knowing this
propagation time and the distance between Z_1 and Z_2, one can calcu-
late the velocity of sound through the sample. However, in the present
technique it is necessary to record the pulse propagation time only;
the velocity conversion is not required.

For dynamic thermoacoustical analysis the sample was heated in
air atmosphere at a programmed rate of 20°C/min. The system
temperature was continuously recorded on a DTA chart whereas the
sonic response was simultaneously recorded on a time base recorder
provided with the pulse propagation meter. The abscissa of the
original sonic chart was later converted to a temperature scale.

Simultaneous DTA and Dynamic Thermoacoustical
Analysis

In the case of simultaneous differential thermal analysis and
dynamic thermoacoustical analysis, the fibers were cut into small

pieces by using a Wiley mill with 60 mesh screen, and 5 mg of this sample was poured into a melting point tube. The tube was then placed into the sample cavity T_2; and a thermocouple, as shown in Fig. 1, was inserted into the sample. Similarly, in the reference cavity T_3, a melting point tube containing reference glass beads was inserted. The reference thermocouple was then embedded into the glass beads. For thermoacoustical analysis, the set-up was the same as that described in the preceding section.

On heating the metal block A, the DTA curve of the sample was obtained on the X-Y recorder of the DTA instrument and a thermoacoustical curve of the sample was obtained on the time base recorder of the PPM-5R.

Samples

The following fiber samples were used: nylon 610 mono filament (0.023 mil), nylon 66 filament (70/30), Alrac nylon 4 filament, Vylor mono filament (2750 denier), Dacron polyester filament (70/34), Fortrel polyester filament (70/36), Teflon filament (1200/180), rayon filament (900/100), cotton spun yarn, and glass filament.

Dacron polyester, Teflon, and all nylon filaments, except nylon 4, were manufactured by Du Pont. Nylon 4 was supplied by Alrac Division of Radiation Research Corp. Vylor, which contains nylon 66 and nylon 610, was also manufactured by Du Pont. Fortrel polyester was supplied by Celanese Corp. Rayon fiber was supplied by American Viscose and cotton yarn was obtained from Coats and Clark. The glass fiber was that used in torsional braid apparatus, supplied by Chemical Instruments Corp.

RESULTS AND DISCUSSIONS

Dynamic Thermoacoustical Curves

A sketch of a hypothetical dynamic thermoacoustical curve is shown in Fig. 2. The pulse propagation time for a distance of x cm of the sample at room temperature is represented by the horizontal portion of the curve AB. As long as the distance x is kept constant, AB remains parallel to the abscissa. The velocity of sound through the material at $25°C$ is equal to $(x/120) \times 10^6$ km/sec. The temperature programming of the DTA apparatus is initiated at B. As long as the sample remains physically and chemically unchanged, the curve continues to indicate a horizontal line. At C the sample begins to transform to a different phase and the curve deviates from the base line. A change toward the upward direction indicates the lowering of the sound velocity. It is

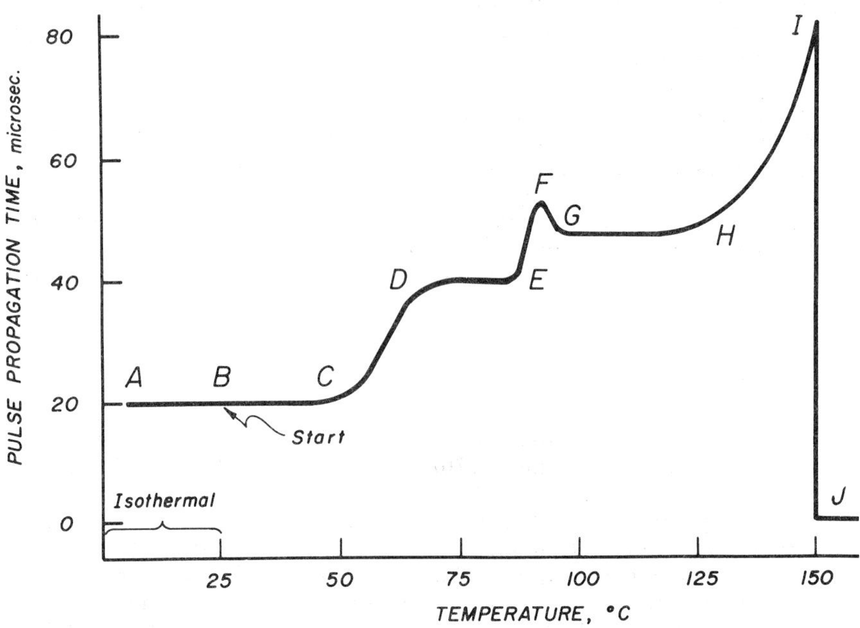

FIG. 2. A hypothetical dynamic thermoacoustical curve.

known that the velocity of sound is highest in solid, lowest in gas, and intermediate in liquid. Therefore, one may assume that the upward trend of the curve would indicate the change of polymer from compact form to relatively fluid form or, in other words, an increase of molecular motion of polymers. An opposite phenomenon is indicated by the downward trend of the curve FG. The sample at G is certainly in a less fluid state than at F. Again G to H shows no physical or chemical change in the sample. At H the sample reveals the premelting behavior. As the melting starts there is a sharp upward trend of the curve until the sample breaks at I due to the actual melting. The recorder pen drops immediately to zero, indicating thereby a discontinuity of the pulse propagation path.

In the case of a metal wire, the interpretation of the thermoacoustical curve is much simpler. For a metal, the sonic velocity is related to Young's modulus according to

$$C = (Y/D)^{1/2}$$

where C is the sonic velocity, D is the density of the metal, and Y is the Young's modulus. Therefore, the entire curve can be interpreted as the

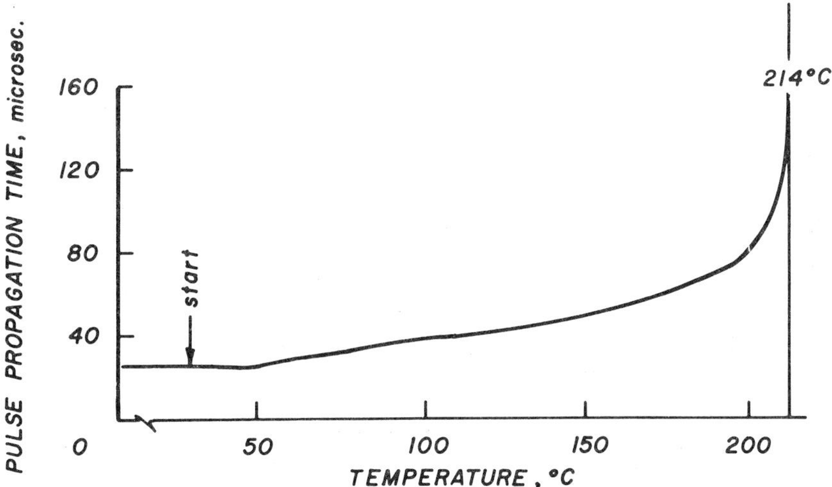

FIG. 3. Dynamic thermoacoustical curve of nylon 610 (draw ratio 3.5 to 1) in air.

change of Young's modulus or the density. In the case of polymers, the interpretation is not as simple due to the viscoelastic behavior of the material. However, before entering into any theoretical discussion it should be more interesting to review some of the experimental curves of synthetic and natural fibers.

The thermoacoustical curve of nylon 610 filament drawn to a ratio of 3.5 to 1 is shown in Fig. 3. The beginning of the temperature programming is indicated in the figure by an arrow. The curve shows a distinct upward deviation from the baseline at 45°C. It levels off again at 55°C. This deviation is attributed to the glass transition temperature of nylon 610. Premelting behavior of the polymer is revealed by the upward trend of the curve above 150°C. The melting is indicated by the sharp upward trend of the curve and then an instantaneous drop to the zero line. It is important to note that above the glass transition temperature the pulse propagation time increased continuously with the rise of temperature. Prior to melting, however, the propagation time increased at an accelerating rate.

The behavior of undrawn nylon 610 and drawn nylon 610 at different draw ratios has been compared in Fig. 4. The increase of pulse propagation time near T_g is more pronounced in the case of the undrawn fiber. The fiber drawn to a ratio of 3.5 to 1 shows relatively insignificant change at the time scale indicated in the figure. Another deviation in the curve of undrawn nylon 610 was observed at 160°C.

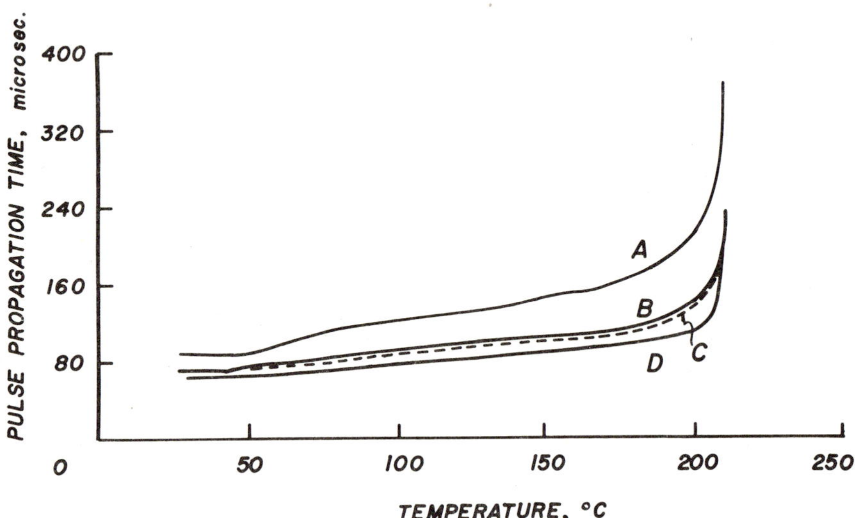

FIG. 4. Dynamic thermoacoustical curves of undrawn and drawn nylon 610 in air. (A) Undrawn. (B) Draw ratio 2:1. (C) Draw ratio 2.5:1. (D) Draw ratio 3.5:1.

This change may be associated with certain crystal structural changes. The premelting behavior of undrawn and drawn fibers appeared to be similar and melting of the polymer occurred at the same temperature in all cases. It should be noted that the curves shifted downward as the draw ratio was increased. This shift toward the lower pulse propagation time or, in other words, toward the higher sonic velocity has been attributed to the increase of the degree of orientation of the polymer molecules with the increased draw ratio. A theoretical consideration on this aspect is given later in this paper.

Thermoacoustical curves of a variety of synthetic fibers are shown in Fig. 5. These curves were all obtained under the same condition and at the same scale sensitivity. The curves of nylon 4, nylon 66, and nylon 610 all show different characteristic natures. Dacron and Fortrel polyester fibers behaved differently at temperatures below 100°C, but above 100°C both of them behaved identically. By comparing the curves, the thermoacoustic behavior of polyester fibers seemed to be somewhat similar to nylon 66. Teflon behaved similar to nylon 4 between 100 and 175°C. However, above 175°C it continued to rise upward at the same rate whereas the nylon 4 curve shows a change in rate. Also, Teflon melted at a lower temperature than nylon 4.

The thermoacoustical behavior of a glass fiber indicates no

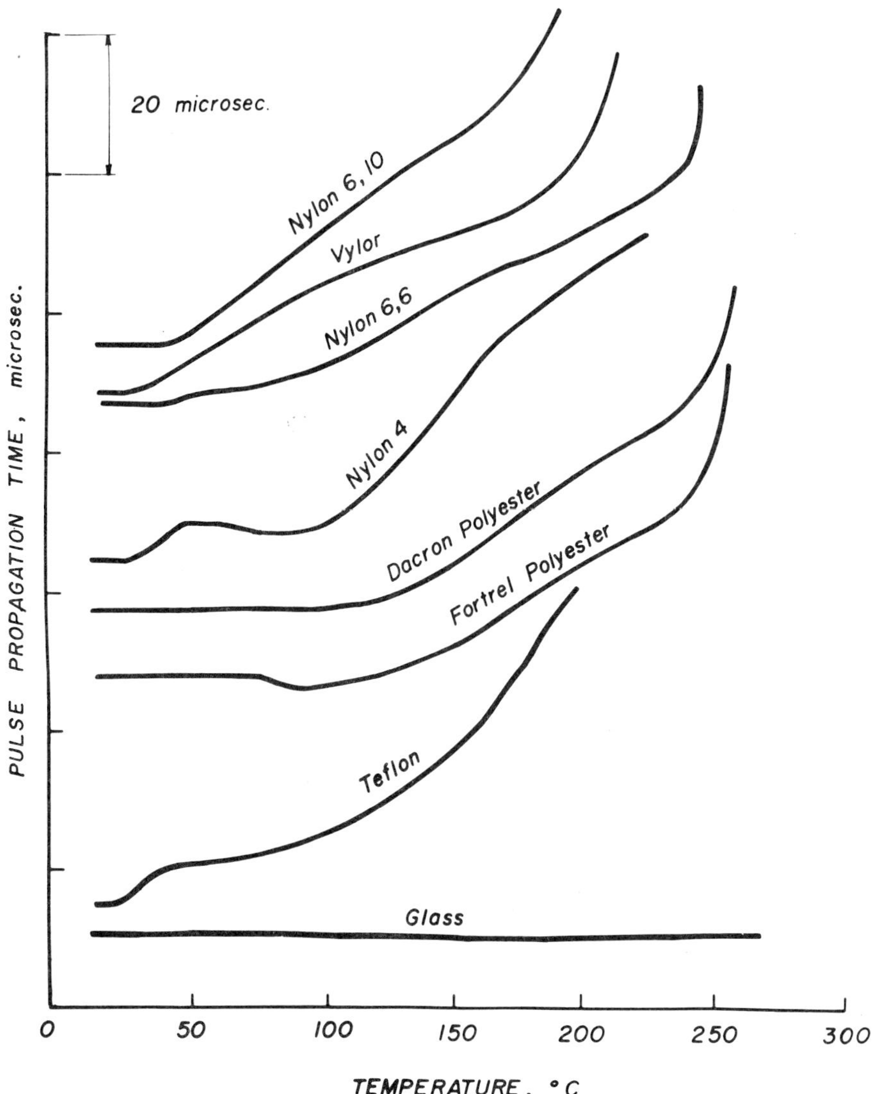

FIG. 5. Dynamic thermoacoustical curves of synthetic fibers in air.

significant change in the temperature range shown. The horizontal
straight curve for glass fiber further indicates that the curves for
synthetic fibers are actually showing the characteristic features of
the fibers and are not just experimental artifacts.

Cellulose fibers, such as cotton and rayon, do not melt but decompose at temperatures above 300°C. Cotton and rayon are
chemically the same (polymer made up of cellobiose units) but
differ in their morphological structure, degree of crystallinity, and
degree of polymerization. They are hard to differentiate by differential thermal analysis or by thermogravimetric analysis. The
dynamic thermoacoustical curves of cotton and rayon are shown in
Fig. 6. They are distinctly different, particularly above 250°C.
Rayon shows a distinct peak at about 340°C, whereas cotton shows
a series of overlapping peaks at higher temperatures. These peaks
can be attributed to the decomposition of cellulose fibers. Because
of a variety of chemical changes at the decomposition temperature,
such as polymer scission and end-group unzipping, the velocity of
pulses was slowed down, resulting in a peak. The right-hand side of
the peak indicates the resumption of the original speed as the
chemical changes were over and the cellulose molecule was converted
to a stable carbonized form. Again, the curve of glass fiber has been
included as a reference.

A simultaneous thermoacoustical curve and differential thermal
analysis curve was obtained with nylon 610 by the technique described
in the experimental section. For the thermoacoustical curve the
sample was mounted as shown in Fig. 1, and for DTA the sample was
cut into small pieces. Both curves were obtained simultaneously as
shown in Fig. 7.

Theoretical Interpretation

The dynamic thermoacoustical technique and some of the experimental curves have been discussed. It is now important to investigate
what the thermoacoustical curve reveals and how these sonic pulse
propagation responses can be related to fundamental polymer properties. In order to develop a theory, the Maxwell-Wiechert model [4]
for polymers has been considered as the basis. The model, shown in
Fig. 8, consists of a very large (or infinite) number of Maxwell
elements (spring and dashpot) coupled in parallel. In the figure, E_i
and τ_i are the modulus and relaxation time, respectively, of the i-th
element where i ranges from 1 to m and $\eta_i^{(t)}$ is the viscosity in
tension, which is linearly related to the shear viscosity η. When
strain s is applied to the model, a stress f will be developed. It must
be remembered that the only observables are the external strains s
and the overall stress f. The partial stress f_i and the quantities E_i
and τ_i are not microscopically observable.

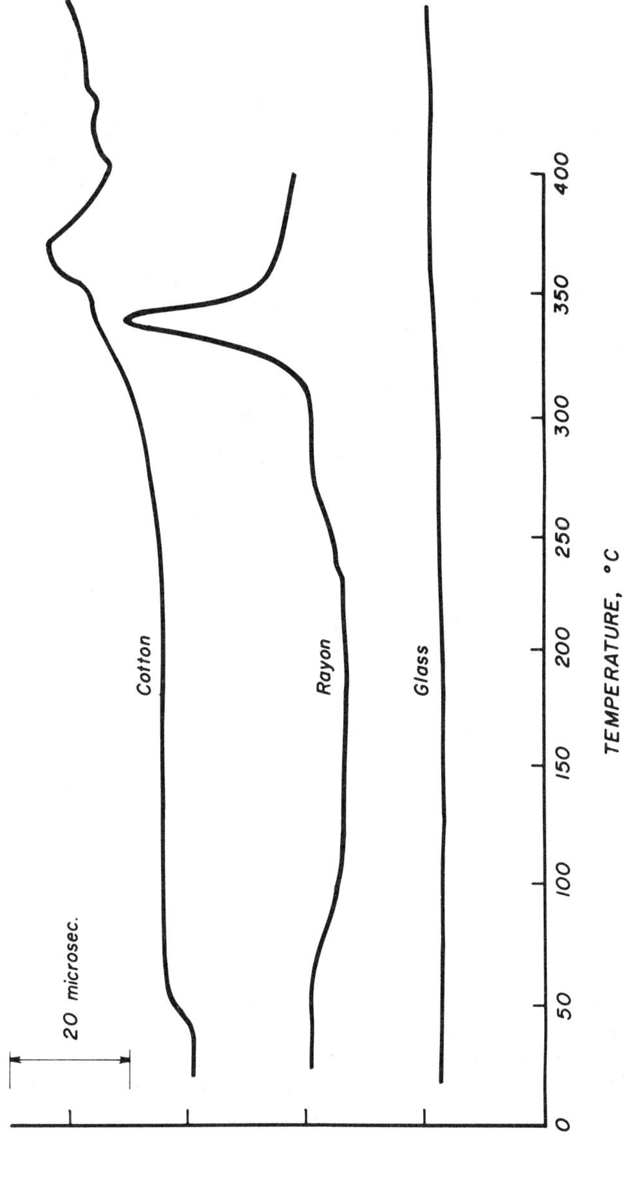

FIG. 6. Dynamic thermoacoustical curves of cellulose fibers in air.

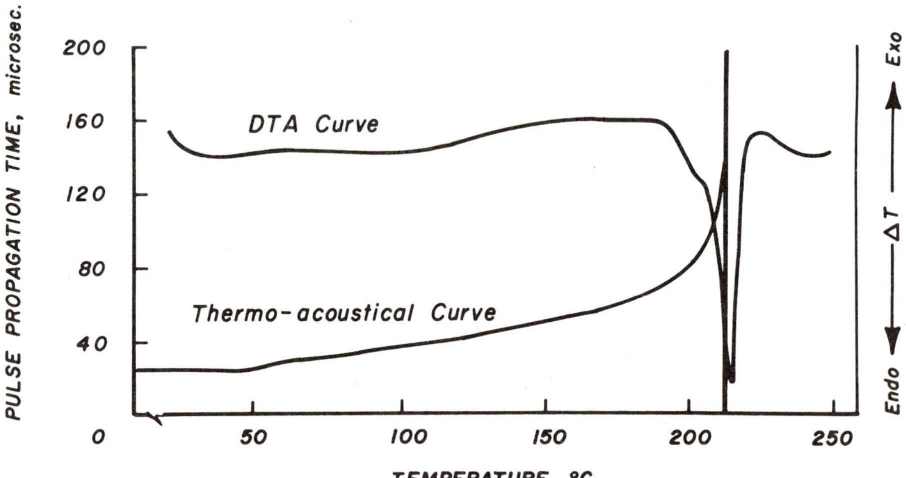

FIG. 7. Simultaneous DTA and dynamic thermoacoustical analysis
of nylon 610 in air.

Assuming that a cube of viscoelastic material (Maxwell-Wiechert
model) of unit dimensions is attached to an external mass and the
whole system is subjected to forced oscillations, an equation can be
deduced to define the viscoelastic stress set-up in the material in
terms of the complex dynamic modulus [5]. Tobolsky and Eyring
[6] separated the real and imaginary parts of the equation and defined the dynamic modulus (E_{dyn}) of the material as follows:

$$E_{dyn} = \sum_{i=1}^{m} \frac{E_i\, \omega^2\, \tau_i^2}{1 + \omega^2\, \tau_i^2} \tag{1}$$

where ω is the frequency of the applied stress.

When sonic pulses of frequency ω are transmitted through a polymeric
material, the dynamic modulus can be defined by the equation cited
above. Nolle [7] deduced a separate relationship between the dynamic
modulus and the velocity of pulse propagation for an unoriented
polymer:

$$E_{dyn} = \rho c^2 \omega^2 (\omega^2 - \alpha^2 c^2)/(\omega^2 + \alpha^2 c^2)^2 \tag{2}$$

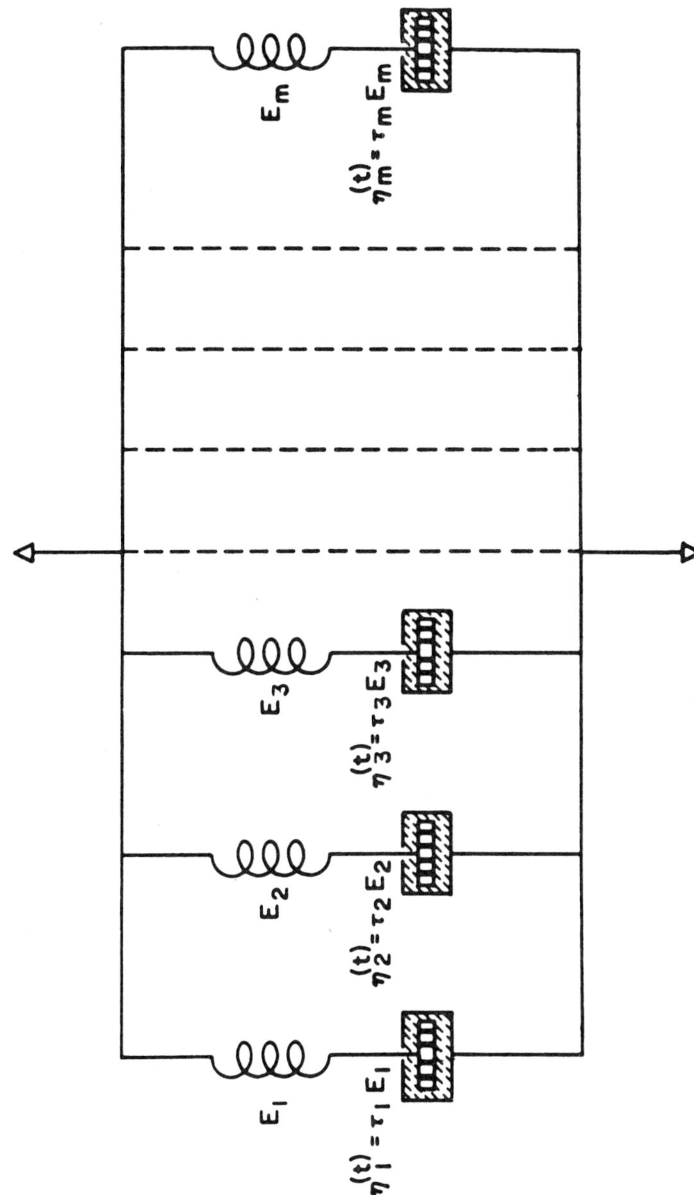

FIG. 8.   Maxwell-Wiechert model for polymers.

where ρ is the density of polymer, c is the velocity of sound, and α is attenuation. For small damping, α is very small and the Eq. (2) reduces to

$$E_{dyn} = \rho c^2 \tag{3}$$

For an unoriented polymer, by combining Tobolsky and Eyring's equation (Eq. 1) with Nolle's equation (Eq. 3) the following equation results:

$$\rho c^2 = \sum_{i=1}^{m} \frac{E_i \omega^2 \tau_i^2}{1 + \omega^2 \tau_i^2} = \sum_{i=1}^{m} \frac{E_i \tau_i^2}{1/(\omega^2 + \tau_i^2)} \tag{4}$$

If, however, the elements of the model have a very high relaxation time (τ_i) and the frequency of stress ω is also very high, Eq. (4) reduces to

$$\rho c^2 = \sum_{i=1}^{m} E_i \tag{5}$$

Since a polymer block contains elements of different τ values ranging from very low to very high values, let us assume that a fraction of elements have τ values so high that their relaxation is not responsive to sonic waves of frequency ω, and the rest of the elements have relatively lower τ values and the relaxation of these elements is responsive to sonic waves. In the former case the sound waves will propagate according to Eq. (5) whereas in the latter case the waves will propagate according to Eq. (4). Let us assign the latter fraction as β and the former as $(1 - \beta)$. The propagation of sound pulses through a polymer consisting of these two broadly classified groups of elements is illustrated in Fig. 9. The elements having low τ values are represented by open circles and their relaxation time is designated by τ_i, and those of high τ values are represented by solid circles and their relaxation time is designated as τ_j. The velocity of sound will be influenced by the combined effect of all these elements. Assuming that they are randomly distributed, the ultimate sound velocity through the polymer can be obtained by applying Urick's approximation [8]. The resultant equation is

$$\frac{1}{\rho c^2} = \frac{\beta}{\displaystyle\sum_{i=1}^{m\beta} \frac{E_i \omega^2 \tau_i^2}{1 + \omega^2 \tau_i^2}} + \frac{1 - \beta}{\displaystyle\sum_{j=1}^{m(1-\beta)} E_j} \tag{6}$$

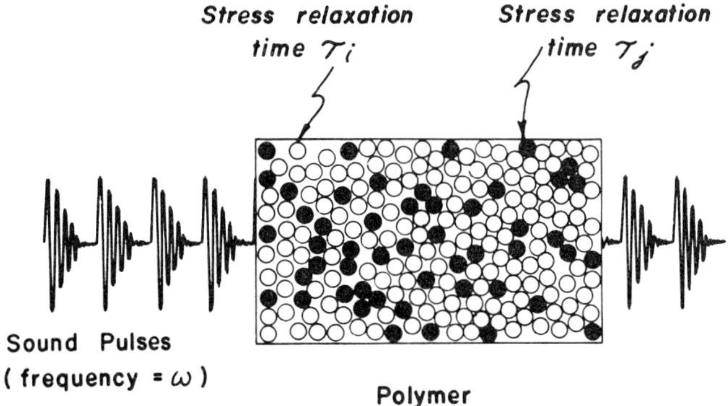

FIG. 9. A schematic showing pulse propagation through a polymer consisting of broadly classified elements ($\circ$) Elements of low relaxation time, τ_i. ($\bullet$) Elements of high relaxation time, τ_j.

where subscripts i and j represent the elements of low and high τ values, respectively. This equation is applicable to nonoriented polymers only.

For oriented polymer molecules Moseley [3] derived a relationship between velocity of sound and molecular orientation:

$$c_\theta = \frac{2}{3} \frac{c^2}{\sin^2 \theta} \tag{7}$$

where θ is the angle between the direction of the propagation of sound and the molecular axis as shown in Fig. 10. c_θ is the velocity of sound in the direction as shown in Fig. 8, and c is the velocity in the same nonoriented polymer.

Therefore, for oriented polymer molecules that exist in fibers, Eqs. (6) and (7) can be combined. Hence

$$\frac{1}{c_\theta^2} = 0.67 \sin^2 \theta \left[\frac{\rho \beta}{\displaystyle\sum_{i=1}^{m\beta} E_i \sin^2 \delta_i} + \frac{\rho(1-\beta)}{\displaystyle\sum_{j=1}^{m(1-\beta)} E_j} \right] \tag{8}$$

where δ_i is the phase angle between the stress and the strain and

$$\tan \delta_i = \omega \tau_i$$

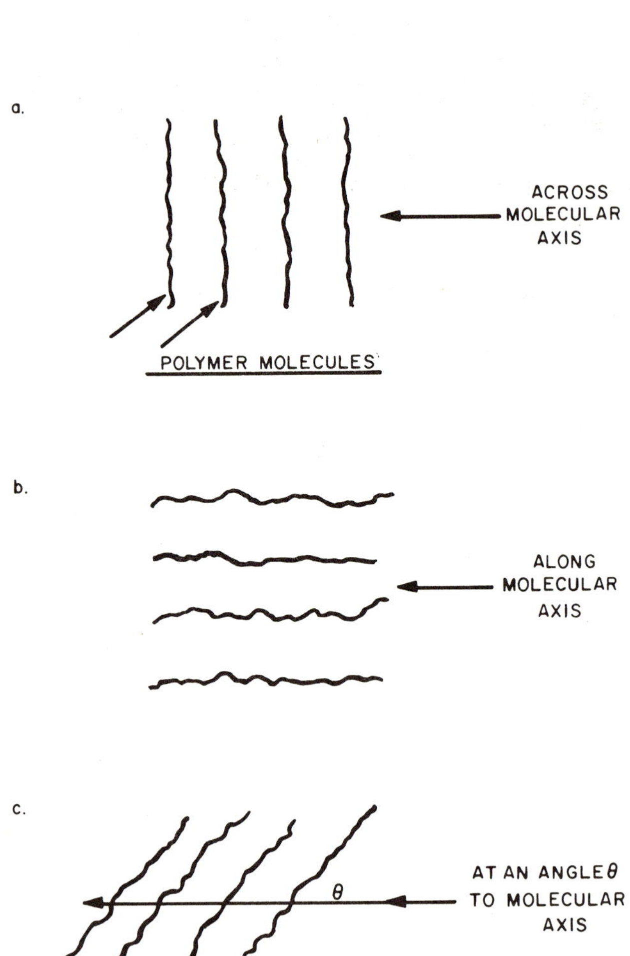

FIG. 10. Possible modes of sound transmission in polymers [3].

By changing c_θ to pulse propagation time, μ, for a fixed length, χ, of the sample, we get

$$\mu = 0.82\chi \sin\theta \left[\frac{\rho\beta}{\sum_{i=1}^{m\beta} E_i \sin^2 \delta_i} + \frac{\rho(1-\beta)}{\sum_{j=1}^{m(1-\beta)} E_j} \right]^{1/2} \qquad (9)$$

This is the final relationship between the pulse propagation time and the viscoelastic parameters of a polymer. In Figs. 2 to 6, all the curves represent μ, which is a function of orientation and viscoelastic parameters vs temperature. For simplicity and practical consideration, let us write the equation as

$$\mu = 0.82\chi \sin\theta\, \epsilon_s \qquad (10)$$

where ϵ_s is defined as a sonic viscoelastic function of the polymer at a constant sonic frequency ω, and this function is equal to

$$\left[\frac{\rho\beta}{\sum_{i=1}^{m\beta} E_i \sin^2 \delta_i} + \frac{\rho(1-\beta)}{\sum_{j=1}^{m(1-\beta)} E_j} \right]^{1/2}$$

Thus it can be inferred from Eq. (10) that as the fiber sample is heated at a constant rate, the thermoacoustical curves will be influenced by 1) the change in orientation of the polymer molecules and 2) the change in the viscoelastic properties of the polymer.

Experiments were conducted with repeated heating and cooling of nylon 610 to investigate the effect of heating on the orientation of polymer molecules. Figure 11, Curve A, represents the first heating of nylon 610 filament. The sample was allowed to cool slowly and found to indicate the same pulse propagation time μ at room temperature as that of the original sample. The second heating is represented by Curve B. On cooling after the second heating the propagation time was back to the original value. A similar phenomenon was also observed on heating a third time and cooling at room temperature. Since the sample was not mechanically disturbed during the heating and cooling process, it was expected that any changes in the orientation during heating would have shown some residual effects even after cooling the sample to room temperature. Any changes in the orientation should have been reflected as a displacement of the curve at room temperature after the second

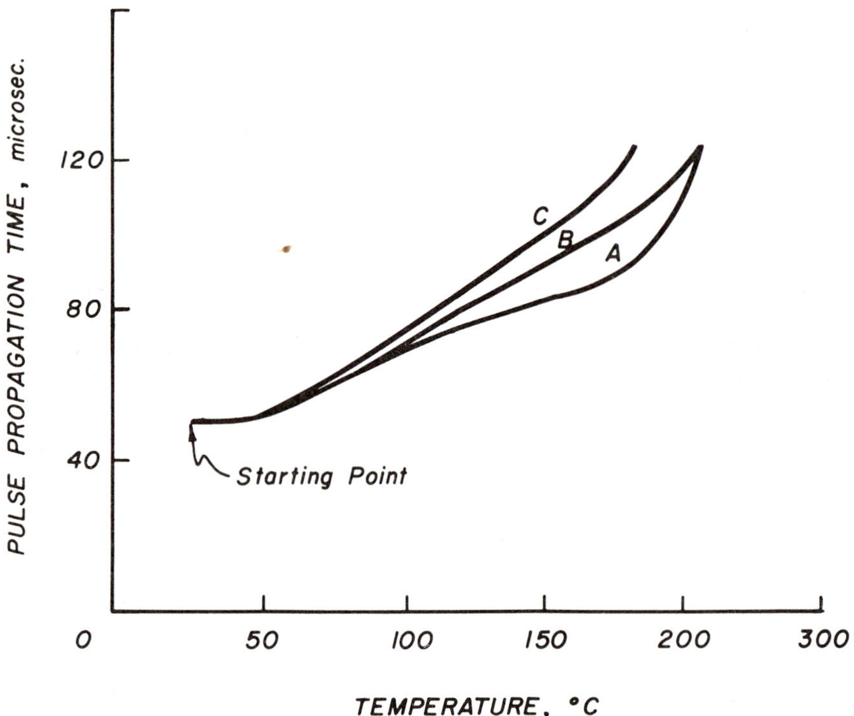

FIG. 11. Dynamic thermoacoustical curves of nylon 610 under repeated heating. (A) First heating. (B) Second heating. (C) Third
heating.

and third cooling (refer to Fig. 4 also). Experimental curves show no
such displacement. The fact that the first, second, and third heating
curves do not coincide with each other at higher temperatures can be
attributed to certain irreversible changes in the viscoelastic properties of the polymer. These experiments strongly suggest that the
orientation of the polymer molecules in the fiber does not undergo any
significant change under these experimental conditions. Therefore, it
can be concluded that the sonic response under dynamic heating
condition would enable us to characterize the fibers in terms of their
viscoelastic responses to heat.

CONCLUDING REMARKS

The main purpose of this work is to demonstrate that acoustical
measurement under dynamic heating conditions can be used for

characterizing certain physicochemical properties of polymers. A theorectical relationship was derived between the sound propagation time, the viscoelastic properties of polymers, and the orientation of molecules. This relationship should be useful for the interpretation of dynamic thermoacoustical curves of fibers and other polymers.

However, further development is required in the experimental technique, particularly in the instrumentation. More precise measurements can be made by inserting the entire fiber mounting set-up and both piezoelectric transducers inside the heating chamber. In order to do this, selection of appropriate transducers which could stand temperature up to $600/700°C$ is important. Atmospheric control and an ability to cool the furnace below room temperature are also essential. In brief, to build a standard instrument the experimental set-up shown here must be modified.

The theoretical relationship derived here could be further simplified by separating the orientation factor and viscoelastic function so that they can be analyzed independently. The sonic viscoelastic function could be solved further to determine the individual parameters such as elastic modulus, dynamic viscosity, and stress relaxation time. Many of these proposed modifications can be achieved if the amplitudes of the pulses are also continuously measured along with the velocity.

ACKNOWLEDGMENTS

The author wishes to acknowledge the experimental assistance of Mrs. Dolores Scott. He also thanks the management of Personal Products Company for permission to publish this study.

REFERENCES

[1] J. W. Ballou and J. C. Smith, J. Appl. Phys., 20, 493 (1949).
[2] W. H. Charch and W. W. Moseley, Jr., Text. Res. J., 29, 525 (1959).
[3] W. W. Moseley, Jr., J. Appl. Polym. Sci., 3, 266 (1960).
[4] E. Wiechert, Wied. Ann. Phys., 50, 335, 546 (1893).
[5] A. V. Tobolsky, Properties and Structures of Polymers, Wiley, New York, 1962, p. 115.
[6] A. V. Tobolsky and H. Eyring, J. Chem. Phys., 11, 125 (1943).
[7] A. W. Nolle, J. Acoust. Soc. Amer., 19, 194 (1947).
[8] R. J. Urick, J. Appl. Phys., 18, 983 (1947).

Pyrolysis-Molecular Weight Chromatography of Polymers: A New Technique

ERDOĞAN KIRAN and JOHN K. GILLHAM

Polymer Materials Program
Department of Chemical Engineering
Princeton University
Princeton, New Jersey 08540

ABSTRACT

A pyrolyzer with programming capability has been coupled in series with a thermal conductivity cell and a mass chromatograph. The thermal conductivity cell gives the flexibility for selective trapping of decomposition products, and also provides data that are complementary to thermogravimetric and differential thermal analyses. The mass chromatograph is composed of two gas chromatographs that are run parallel from a common injection port. Each chromatograph uses a different carrier gas and is equipped with a gas density balance detector. The instrument simplifies identification of mixtures of unknowns in directly providing molecular weights, absolute quantities, and gas chromatographic retention times of the constituents. In this respect, the present system is analogous to pyrolysis-gas chromatography-mass spectrometry.
Discussion of the technique and its application to the investigation of the thermal degradation of low-density polyethylene are presented.

INTRODUCTION

Identification of decomposition products is a necessary precursor
to the study of the mechanism of thermal degradation of polymers.
Among the separation techniques, gas chromatography has received
much attention in the past decade; in particular, pyrolysis-gas
chromatography has gained wide acceptance as a technique of polymer
analysis [1-4]. However, positive identification by gas chromatography
is not unambiguous without the use of additional analytical techniques
[5-7] such as IR, mass spectrometry, and NMR.

Molecular weight chromatography (which is not to be confused with
mass spectrometry) is a relatively new technique [8-10]. The commercially available instrument is called a mass chromatograph [11].
The single instrument directly provides molecular weights, absolute
quantities and gas chromatographic retention times of the constituents
of a mixture by using gas density balances as the detectors for the
effluents from gas chromatographic columns. It thus upgrades gas
chromatography and facilitates the process of identification of
unknowns.

The present paper discusses a series combination of a pyrolyzer,
a thermal conductivity cell, and a mass chromatograph as applied to
polymer degradation. The pyrolyzer is designed to carry out decomposition by either program- or flash-heating. The thermal
conductivity cell is included to monitor the formation of volatile
decomposition products and to permit their selective trapping and
subsequent selective desorption from a trap. The mass chromatograph
achieves separation of the decomposition products in the manner of
a gas chromatograph, but also provides information on their molecular
weights with ease (compared with mass spectrometry).

EXPERIMENTAL

Apparatus

A schematic drawing of the present system is shown in Fig. 1. The
pyrolyzer-thermal conductivity probe assembly consists of a tubular
reactor and heater (pyrolyzer), a detector oven (dotted enclosure)
which houses the thermal conductivity cell and the two six-port valves
(X and Y), and an external trap (pyrolyzer trap). This assembly was
custom-designed for this laboratory and in a modified form is now
commercially available [11]. The pyrolysis chamber is a quartz
tube (3/16 in. i.d.) which can be program- or flash-heated to 800°C.
The sample to be pyrolyzed is placed in the (heat-cleaned) quartz

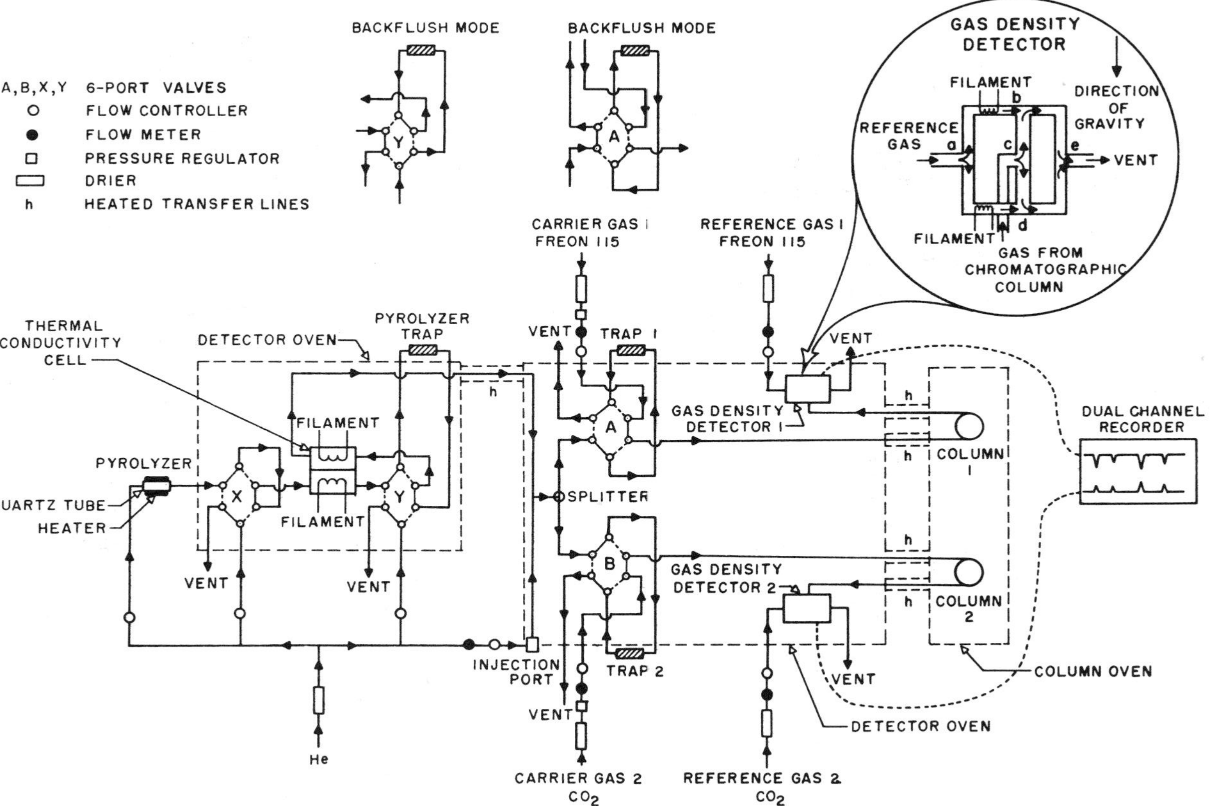

FIG. 1. Pyrolyzer/thermal conductivity probe/mass chromatograph.

tube between two loosely packed (heat-cleaned) glass-wool plugs. The glass wool plugs help position the sample and prevent mechanical loss. The quartz tube, with sample in place, is assembled using O-rings and Swagelok fittings. After assembly, the tube is purged with helium while Valve X is at the vent position (in which mode the dotted lines in the valve are connected and the solid lines are broken). The Valve X is then turned to the sampling mode (as in the figure) and the reactor heater (which surrounds the tube and is movable along it) is properly positioned and activated to perform the decomposition at a selected heating rate. Flexibility for flash pyrolysis has been provided, in which case the furnace, which is movable along the quartz tube, is heated in advance to the desired temperature, and then moved over the sample. The volatile products are transported by helium carrier gas through Valve X and then through one arm of the thermal conductivity cell where they are detected and then either trapped or vented from Valve Y. Trapping of the decomposition products is achieved by having Valve Y in its sampling mode (as shown in the figure) in which position the fragments are carried to a trap situated outside the detector oven. The trap is a stainless-steel column (1/8 in. o.d. by ~16 in.) packed with cross-linked polystyrene beads (Porapak Q, 80-100 mesh, Waters Associates). The trap has the proper configuration so that a fraction of it (~4 in. section) can be placed in a Dewar for subambient cooling. When the decomposition products enter the trap, the high molecular weight constituents are retained by the packing at the entrance of the trap, which is at room temperature. The more volatile constituents are retained by the packing downstream which is in the section cooled by liquid nitrogen. After trapping, Valve Y is turned to the backflush mode which reverses the direction of helium flow (Fig. 1, top left), the Dewar is removed, and the trap is flash-heated to a desired temperature below $250°C$ (the limit of thermal stability of Porapak Q), upon which the trapped constituents are released and carried by helium through Valve Y and then through the other arm of the thermal conductivity cell where they are again detected, this time as a sharp front, and then are delivered through a heated transfer line into the mass chromatograph. If desired, the trap heater can be program-heated. This will result in the fractional release of the trapped constituents. Subsequent detection by the thermal conductivity cell will therefore provide a gas chromatographic output in its own right.

The mass chromatograph was introduced about 3 years ago [8, 11]. It consists of two independent gas chromatographic systems using two different carrier gases which are run parallel from a common injection port. Each system is composed of a six-port valve (A, B), an external trap (Trap 1, Trap 2), a gas density balance detector (Detector 1, Detector 2), and a chromatographic column (Column 1, Column 2). The detectors and the valves are housed in the detector oven. When a sample is introduced to the mass chromatograph, whether directly from the pyrolyzer or through the one injection port, it is carried by helium and split into two approximately equal fractions (splitter). Each fraction is carried by helium through the respective valves into the respective external traps. The traps are similar to the pyrolyzer trap. After trapping, Valves A and B are simultaneously turned to the back-flush positions (shown for Valve A at top center of Fig. 1). Then, carrier gases 1 and 2 flow through the respective traps and, after flash heating of the traps, carry the released fractions into the respective columns where separation is achieved. The constituents are detected as they elute from the matched columns by the respective gas density balance detectors Gow-Mac type, equipped with tungsten wire filaments). The recorder output displays two sets of peaks corresponding to the responses from the two gas chromatographic systems for the same constituents of the mixture. The ratio of responses are then related to the molecular weights of the constituents, as explained in the next section.

Gas Density Balance and Molecular Weight Determination

The introduction of the first gas density balance was motivated by a desire to develop a detector where response would be independent of the chemical structure of the substance being analyzed [12]. It was, however, immediately realized that the detector had an additional unique feature in that it could be used for determination of molecular weights [13, 14]. Later, a simplified gas density detector was developed [15] which formed the basis of the Gow-Mac gas density balance [16].

A schematic of the Gow-Mac type gas density balance detector is shown in Fig. 1 (top right). The reference gas, which is the same as the chromatographic carrier gas, is split at "a" and flows in parallel over the lower and upper filaments which are part of a Wheatstone bridge. If no solute is carried by the carrier gas from the column, the flow of the reference gas over the filaments is not disturbed. However, when a solute with a greater density than the carrier gas enters the cell at c, the density of the gas in the vertical conduit bd becomes greater than that of the pure carrier gas, and consequently the pressure at d increases. This causes the flow of the reference gas to decrease over the lower filament but to increase over the upper filament. The converse phenomenon applies for a solute which has a lower density than that of the carrier gas. The change in the flow rate of the reference gas over the filaments leads to a change in temperature and in turn to a change in the resistance between the upper and the lower filaments. This unbalance is recorded as a gas chromatographic peak.

It should be noted that the flow rate at the exit (vent) remains

constant and that the variation of pressure (at d) leads to a variation
only in the interior flow rates of the gases. The change in density $\Delta\rho$
can be related [15] to the change in flow rate ΔQ by

$$\Delta\rho = \frac{8\mu L\Delta Q}{\pi\, gr^4 h} \tag{1}$$

where μ, L, g, r, and h represent the viscosity of the gas, the total
length of the conduit abcde, the gravitational constant, the radius of the
conduit, and the height of the gas stream in the conduit bd where the
change in density occurs respectively. It should be noted that the
presence of a solute will affect the viscosity of the carrier gas mostly
in the conduit bd. However, the change in viscosity induced by the
introduction of small amounts of solute can be considered to be
negligible, since at a given temperature the viscosities of gases are
of the same magnitude. Even though the validity of the assumptions
(i.e., that incompressible steady-laminar flow conditions prevail,
and that the pressure variation at d leads to identical variation in
flow rate in all conduits of the loop abcda) that are invoked [15] in de-
riving Eq. (1) may be questionable, the proportional relationship
between $\Delta\rho$ and ΔQ has been experimentally verified [15]. Furthermore it has been experimentally shown [15] that the electrical
response of the detector is also linearly related to the change in flow
rate, which therefore is proportional to the change in density. (The
assumptions that lead to Eq. 1 are more suitable for liquids and, in
fact, it has been reported [17] that the gas density balance can be
used as a detector in liquid-solid chromatography.)

To a first approximation, the density of a gas is given by the ideal
gas law expression, and at a given pressure and temperature it is
proportional to its molecular weight:

$$\rho = (P/RT)M \tag{2}$$

[Under standard conditions $\rho = (1/22.4)M$, where ρ is expressed as
grams/liter and M as grams/mole.] The instantaneous average
molecular weight of the carrier gas (M) eluting from the chromatographic column can be expressed as

$$M = (1 - Y)M_c + YM_x \tag{3}$$

where Y is the instantaneous mole fraction of a solute X, and M_x and
M_c are the molecular weights of the solute X and the carrier gas,
respectively. Therefore, since the instantaneous density change is

$$\Delta\rho = \rho - \rho_c$$

where ρ and ρ_c are the densities of the carrier gas with and without the solute, respectively,

$$\Delta\rho = \frac{P}{RT}[M - M_c]$$

$$= \frac{P}{RT}[(1 - Y)M_c + YM_x - M_c]$$

$$= \frac{P}{RT}Y[M_x - M_c] \tag{4}$$

The instantaneous electrical response (E) is given by $E = k\Delta\rho$, and therefore

$$E = k\left(\frac{P}{RT}\right)Y[M_x - M_c] \tag{5}$$

where k is a proportionality constant.

The instantaneous mole fraction of solute is related to the instantaneous number of moles of solute (n_x) and carrier gas (n_c) by

$$Y = \frac{n_x}{n_x + n_c} \tag{6}$$

The total electrical response is evaluated by substitution of this expression into Eq. (5) and integration with respect to time. It is directly proportional to the chromatographic peak area A_x. (In the following derivation, however, this constant of proportionality has not been explicitly introduced since its effect would have been nothing more than a redefinition of k in Eq. 7.)

$$A_x = \int_0^\infty E\,dt$$

$$= \int_0^\infty k\left(\frac{P}{RT}\right)[M_x - M_c]\left[\frac{n_x}{n_x + n_c}\right]dt$$

$$= k\left(\frac{P}{RT}\right)[M_x - M_c]\int_0^\infty \left[\frac{n_x}{n_x + n_c}\right]dt \tag{7}$$

Here it has been assumed that, in the time interval for the elution of solute X, the pressure and temperature of the carrier and solute remain unchanged so that the (P/T) term can be taken outside the integral. For small n_x, one can make the approximation

$$\int_0^\infty \left[\frac{n_x}{n_x + n_c}\right] dt \approx \int_0^\infty \left(\frac{n_x}{n_c}\right) dt \approx \frac{1}{n_c} \int_0^\infty n_x \, dt \tag{8}$$

The integral $\int_0^\infty n_x \, dt$ is equal to the total number of moles of solute X that has eluted, i.e., N_x. Equation (7) now becomes

$$A_x = k\left(\frac{P}{RT}\right)\frac{N_x}{n_c}(M_x - M_c)$$

or

$$= k'\left(\frac{P}{RT}\right)N_x(M_x - M_c) \tag{9}$$

or

$$= k'\left(\frac{P}{RT}\right)W_x\left(1 - \frac{M_c}{M_x}\right) \tag{10}$$

where k' $(=k/n_c)$ is a new constant, and $W_x(=N_x M_x)$ is the total mass of solute that has eluted.

Equations (9) and (10) show that at a given temperature and pressure $k'(P/RT)$ becomes a cell constant for a given carrier gas and can be evaluated from a measurement of the peak area due to a known number of moles, N_x (or known mass, W_x), of a solute of known molecular weight, M_x. The molecular weight of an unknown is then evaluated from a measurement of the peak area due to a known mass or number of moles of the unknown [13, 14]. If, however, a mixture of an unknown and a standard is chromatographed using two different carrier gases, the measurement of the masses (or number of moles) of the unknown and of the calibration compound is no longer required [13, 18] (see below).

If a sample of a mixture containing the unknown of molecular weight M_x and a standard of molecular weight M_s is chromatographed using a carrier gas of molecular weight M_{c_1}, the following equations can be written for the detector response for the standard (A_s) and the unknown (A_x):

$$A_s = k'\left(\frac{P}{RT}\right)(mW)\left(1 - \frac{M_{c_1}}{M_s}\right) \tag{11}$$

$$A_X = k' \left(\frac{P}{RT}\right)(nW)\left(1 - \frac{M_{c_1}}{M_X}\right) \qquad (12)$$

where m and n are the weight fractions of standard and unknown, respectively, in the total amount of sample (W) injected.

Division of Eq. (11) by Eq. (12) leads to

$$\left[\frac{A_s}{A_X}\right]_1 = \left(\frac{m}{n}\right)\left[\frac{M_s - M_{c_1}}{M_X - M_{c_1}}\right]\left[\frac{M_X}{M_s}\right] \qquad (13)$$

Therefore, if the ratio of the weight fractions of the unknown and standard is known, which in practice usually necessitates knowing the weight fraction of each, then M_X can be calculated from the known molecular weights and the ratio of the areas of the chromatographic peaks.

If, however, another sample of the same mixture is chromatographed using a different carrier gas of molecular weight M_{c_2}, one can similarly derive a new expression:

$$\left[\frac{A_s}{A_X}\right]_2 = \left(\frac{m}{n}\right)\left[\frac{M_s - M_{c_2}}{M_X - M_{c_2}}\right]\left[\frac{M_X}{M_s}\right] \qquad (14)$$

Therefore, from division of Eq. (13) by Eq. (14),

$$\frac{\left[\dfrac{A_s}{A_X}\right]_1}{\left[\dfrac{A_s}{A_X}\right]_2} = \left[\frac{M_s - M_{c_1}}{M_s - M_{c_2}}\right]\left[\frac{M_X - M_{c_2}}{M_X - M_{c_1}}\right] \qquad (15)$$

Equation (15) does not involve terms related to the masses of the standard or the unknown. Since molecular weights of standard and carrier gases are known, and since the response ratios can be measured from the chromatographic outputs, M_X can be readily calculated.

It has also been shown that by using several carrier gases with molecular weights greater and less than the molecular weight of the unknown, the molecular weight of the unknown is accurately bracketed and interpolated [18].

This concept of using more than one carrier gas forms the basis of
molecular weight chromatography. However, in the mass chromatog-
graph, instead of chromatographing two samples of an unknown mixture
containing a standard, the first with one carrier gas and the second
with another, a sample (which does not require inclusion of a standard
once the instrument has been calibrated) is injected directly and split
automatically into two fractions (αW and βW, where $\alpha + \beta = 1.0$, and
W = amount injected), and each fraction is simulataneously analyzed
by the respective chromatographic system. Therefore, for a given
solute X, two simultaneous responses, describable by two equations
of the form of Eq. (10), are obtained:

$$A_1 = k_1' \left(\frac{P}{RT}\right)_1 (\alpha W)\left[1 - \frac{M_{c_1}}{M_X}\right] \tag{16}$$

and

$$A_2 = k_2' \left(\frac{P}{RT}\right)_2 (\beta W)\left[1 - \frac{M_{c_2}}{M_X}\right] \tag{17}$$

where A_1 and A_2 are the chromatographic peak areas for unknown X,
k_1' and k_2' are detector constants, and M_{c_1} and M_{c_2} are the molecular
weights of the carrier gases 1 and 2.

Division of Eq. (16) by Eq. (17) leads to

$$K\left[\frac{A_1}{A_2}\right] = \left[\frac{M_X - M_{c_1}}{M_X - M_{c_2}}\right] \tag{18}$$

where

$$K = \frac{\beta k_2'}{\alpha k_1'}$$

(Since both columns and detectors are operated under identical temperature and pressure conditions, P/RT terms cancel.) The knowledge of
the actual value of the split ratio (β/α) is not required, provided that it
does not vary from one experiment to the next. If this condition is
satisfied, K becomes an instrument constant which can be determined by
using a compound of known molecular weight and measuring the response
ratio A_1/A_2 from the chromatographic output. Once K is established,

the molecular weight of an unknown can be determined from Eq. (19) which is obtained by rearranging Eq. (18):

$$M_x = \frac{[(A_1/A_2)K \, M_{c_2} - M_{c_1}]}{[(A_1/A_2)K - 1]} \tag{19}$$

It is clear from Eq. (10) that the detector response is positive if $M_x > M_c$, negative if $M_x < M_c$, and zero if $M_x = M_c$. Some limiting conditions are observed if $M_x \gg M_c$ or $M_x \ll M_c$. When $M_x \gg M_c$, then $M_c/M_x \ll 1$, and $A_x \sim W_x$; that is, the detector response becomes proportional to the mass of the sample and insensitive to its molecular weight. When, however, $M_x \ll M_c$, then $M_c/M_x \gg 1$, and $A_x \sim W_x(M_c/M_x) \sim N_x$; that is, the response becomes proportional to and more sensitive to the number of moles of the sample. Therefore any given carrier gas is expected to display optimal functionality only in a certain molecular weight range. In the mass chromatograph, whereas the use of two carrier gases makes it possible to write Eqs. (16) and (17), and consequently obtain Eqs. (18) and (19), the selection of a high and a low molecular weight carrier gas provides a wide range of operation for calculations of molecular weight. However, selection of the low molecular weight carrier gas requires special considerations. Light gases such as helium or hydrogen should not normally be chosen. This is because, as can be seen from Eq. (10), the detector response is proportional to $(M_x - M_c)/M_x$ and will tend to a constant very rapidly as M_x increases, thus resulting in a narrow molecular weight range of functionality for the detector. Furthermore, with light gases back-diffusion in the detector conduits may become appreciable.

The accuracy of molecular weight determinations depends upon invariance of the instrument constant K, and the accuracy with which the response ratios are measured (see Eq. 19). The relative errors in K and A_1/A_2 are not linearly related to the error they cause in the molecular weight, as is illustrated in Table 1. In the preparation of Table 1, the theoretical value of the factor $K(A_1/A_2)$ was evaluated from Eq. (18) for carrier gases carbon dioxide and monochloropentafluoroethane. Then a defined error was purposefully introduced to the $K(A_1/A_2)$ value, and the molecular weight corresponding to this new value was back-calculated and compared with the initial molecular weight. The errors that are reflected in the molecular weight due to 1 and 5% errors in the ratio $K(A_1/A_2)$ are shown (Table 1). Error in terms of mass units becomes magnified in the high molecular weight region. There is an optimum range of molecular weights for operation.

TABLE 1. Influence of Error Made in Evaluation of $K\left(\dfrac{A_1}{A_2}\right)$ on the Calculation of Molecular Weights

$M_x \left[K\dfrac{A_1}{A_2} = \dfrac{M_x - M_{c_1}}{M_x - M_{c_2}}\right]$	$M'_x \left[= \dfrac{\xi M_{c_2} - M_{c_1}}{\xi - 1.0}\right]^b$	% error $\left[\dfrac{M_x - M'_x}{M_x} \times 100\right]$	$M''_x \left[= \dfrac{\xi M_{c_2} - M_{c_1}}{\xi - 1.0}\right]^c$	% error $\left[\dfrac{M''_x - M_x}{M_x} \times 100\right]$
4 3.76	4.53	13.25	6.54	63.50
10 4.25	10.42	4.20	12.11	21.10
20 5.60	20.28	1.40	21.37	6.85
30 8.88	30.15	0.50	30.74	2.47
40 28.54	40.04	0.10	40.19	0.47
50 17.44	49.94	0.12	49.73	0.54
60 5.91	59.85	0.25	59.33	1.11
70 3.25	69.80	0.28	69.04	1.37
80 2.07	79.76	0.30	78.81	1.48
90 1.40	89.84	0.18	88.73	1.41
100 0.97	99.70	0.30	98.67	1.01
110 0.67	109.70	0.27	108.68	1.20
120 0.45	119.79	0.17	118.85	0.95
130 0.28	129.83	0.13	129.02	0.75
140 0.15	139.85	0.11	139.34	0.47
150 0.04	149.96	0.02	149.79	0.14

160	0.04	160.04	0.02	160.27	0.17
170	0.12	170.09	0.05	170.81	0.47
180	0.19	180.36	0.20	181.62	0.90
190	0.24	190.30	0.15	192.25	1.18
200	0.29	200.67	0.33	204.00	2.00
210	0.33	210.60	0.28	214.12	1.96
220	0.37	220.73	0.33	225.36	2.43
230	0.40	231.21	0.53	236.53	2.84
240	0.43	241.38	0.57	247.71	3.21
250	0.46	251.85	0.74	259.39	3.75
260	0.49	261.86	0.72	270.52	4.04
270	0.51	272.20	0.81	282.30	4.39
280	0.53	282.56	0.91	294.23	5.08
290	0.55	292.77	0.95	306.04	5.53
300	0.57	303.09	1.03	317.67	5.89
320	0.59	323.63	1.13	341.63	6.76
340	0.62	344.96	1.46	367.34	8.04
360	0.65	365.08	1.41	391.88	8.85
380	0.67	386.70	1.76	417.91	9.97
400	0.69	407.33	1.83	443.31	10.82
450	0.73	460.80	2.40	512.81	13.95
500	0.75	512.01	2.40	582.39	16.47

[a] M_{c_1} = 154.46 (molecular weight of Freon-115). M_{c_2} = 44.01 (molecular weight of carbon dioxide).

[b] $\xi = K \dfrac{A_1}{A_2} + 0.01K \dfrac{A_1}{A_2}.$

[c] $\zeta = K \dfrac{A_1}{A_2} + 0.05K \dfrac{A_1}{A_2}.$

Carrier and Reference Gases

Experimental studies on the response of the gas density balance as a function of the nature of the carrier gas, flow rates, and density changes have been reported [18-22]. In particular, hydrogen [12, 14, 21], helium [21, 22], nitrogen [12-14, 18, 19, 21, 22], argon [18, 19, 22], carbon dioxide [18, 19, 21], chloromethane [21], difluoroethane [18], dichloro-difluoromethane [18, 21], sulfur hexafluoride [20-22], bromotrifluoro-methane [21], and octafluorocyclobutane [18] have been investigated. The manufacturers of the mass chromatograph have recommended carbon dioxide and sulfur hexafluoride as the low and high molecular weight carrier gases [8, 9, 11].

In this laboratory, carbon dioxide (Matheson, Coleman instrument grade, with a minimum purity of 99.99%) is used as the low molecular weight carrier (and reference) gas, but monochloropentafluoroethane ClC_2F_5 (Freon-115, Du Pont food grade) is used as the high molecular weight carrier (and reference) gas. The fluorocarbon (bp -38.72°C) is shipped in cylinders from which it can be drawn in liquid form at a higher purity level than the designated 99.98 volume-based percent. Its molecular weight is 8.46 mass units higher than sulfur hexafluoride and extends the upper molecular weight limit of sensitivity of the mass chromatograph.

Helium (Matheson, ultra high purity grade, with minimum purity of 99.999% on a neon-free basis) is used as the carrier gas for the pyrolyzer and the mass chromatograph injection port (Fig. 1). [The commercial mass chromatograph has been slightly modified to accommodate helium (instead of carbon dioxide) as the injection port carrier gas.] Use of the same gas at identical flow rates insures that calibration and in-use conditions are the same. Furthermore, the use of helium as the injection port carrier gas gives the flexibility of using liquid nitrogen-cooling with the mass chromatograph traps when in the sampling mode.

An independent cylinder is used for each carrier and reference gas. Freon-115 is vaporized before entry into the instrument. Room temperature fluctuations are kept at a minimum to prevent condensation and re-evaporation which upset the detector response. Since the response mechanism of the gas density balance involves pressure variation in the cell, any sudden external variations in pressure are immediately detected; this becomes especially important with the Freon-115 reference gas for which flow rates are low.

Helium (carrier) and Freon-115 (carrier and reference) are dried by passing through a column of P_2O_5/firebrick, and then through a column of molecular sieves (Linde, type 5A). Carbon dioxide (carrier and reference) is dried by passing through a column of silica gel, and then through a column of P_2O_5/firebrick. The reason for high purity requirements and extensive drying of gases is to prevent large baseline drifts that become a special problem with the detector using the high molecular weight carrier gas, due to its high sensitivity to low molecular weight constituents (see Eq. 10). In this connection, elimination of leaks in the system becomes a strict requirement. It was observed that the 6-port valves can suffer from leaks, and meticulous care must be taken to eliminate these if proper operation is to be achieved.

Operational Conditions

The gas flow rates were set as follows: 20 ml/min at 50 psig for helium flow; 10 ml/min at 100 psig for carbon dioxide and Freon-115 carrier gases; 41 ml/min at 56 psig for Freon-115 reference gas, and 122 ml/min at 100 psig for carbon dioxide reference gas.

The thermal conductivity and gas density detector currents were set at 100 mA. The temperatures of the thermal conductivity and gas density detector ovens were maintained at 230 and 242°C, respectively. The pyrolyzer and the mass chromatograph trap temperature limits were set at 230 and 235°C, respectively. (Thus the constituents that are released from the pyrolyzer trap are not later retained in the mass chromatograph traps.) The transfer line temperature from the pyrolyzer to the mass chromatograph was maintained at 300°C.

Chromatographic Columns

Two stainless steel (1/8 in. × 12 ft) matched columns packed with 10% silicone gum SE-30 on 60/80 mesh Chromosorb W-AW were used for analysis of the degradation products of hydrocarbon polymers.

Instrument Calibration

The process of instrument calibration involves the determination of the constant K in Eq. (18). In principle, K can be evaluated by chromatographing just one solute of known molecular weight and measuring the response ratio A_1/A_2 from the chromatographic output. However the accuracy of measuring the response ratio varies with molecular weight and therefore, in practice, K should be evaluated using a mixture of known compounds.

Furthermore, in practice, more accurate values of molecular weights of unknowns can be obtained by using more than one value for K, according to the molecular weight range of interest. Accurate measurements of response ratios are difficult to determine in the vicinity of molecular weights of the carrier gases since, as $M_x \rightarrow M_c$, $A_x \rightarrow 0$, and change in response with molecular weight becomes small.

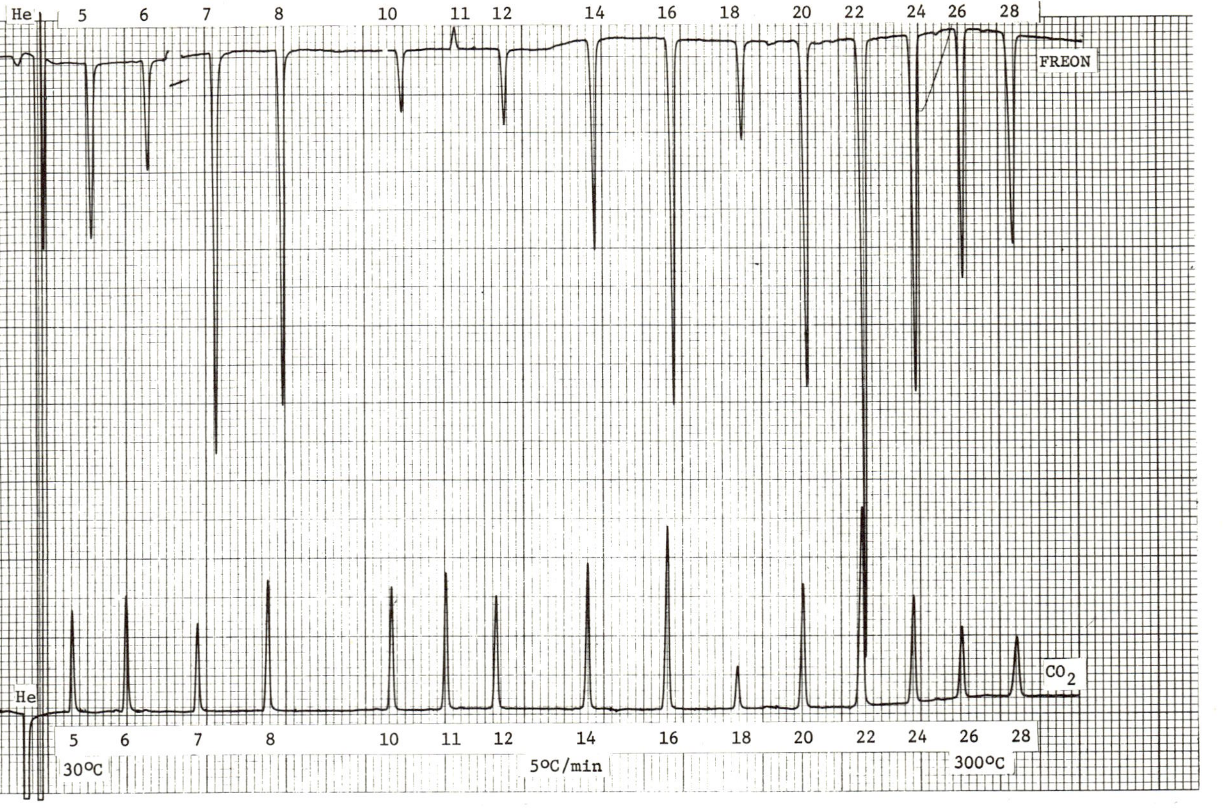

FIG. 2. Mass chromatogram of a mixture of known saturated normal hydrocarbons (C_5 to C_{28}). The chromatogram was obtained by operating isothermally at 30°C until helium peaks eluted, and then by temperature-programming the columns from 30 to 300°C at 5°C/min. The columns were then held isothermally at 300°C. The peak attenuations were ×8 except for the C_5 to C_6 peaks in the Freon-115 channel for which the settings were ×64. After elution of the C_{11} peak, the polarity of the Freon-115 detector response was reversed to result in downward peaks for constituents with molecular weights greater than the molecular weight of Freon-115.

The chromatogram of a mixture of known saturated normal hydrocarbons (C_5 to C_{28}) is shown in Fig. 2 where peaks have been identified with respect to their carbon numbers. The top and bottom series of peaks correspond to the detector responses using Freon-115 and CO_2 carrier gases, respectively.

The instrument constant K was evaluated to be, using the ratios of peak heights, 0.189 in the molecular weight range above the molecular weight of Freon-115 (154.46) and 0.206 in the molecular weight range below the molecular weight of Freon-115. (Due to the intrinsic inaccuracies in measurement as $M \rightarrow M_c$, the molecular weight region $154.46 \pm \sim 30$ was not used in the evaluation of the instrument constant.) Back-calculation of the respective molecular weights of the constituents using the respective K values estimated their molecular weights with, on the average, less than 1.5% and less than 0.5% errors in the respective regions. Calculation of the molecular weights in the region $154.46 \pm \sim 30$ is not affected appreciably by which K value is used since in this region A_1/A_2 is very small and KA_1/A_2 is not sensitive to changes in K. This fact and the difficulty of measurement of A_1 (and thus A_1/A_2) with accuracy in this region justify not considering the compounds with molecular weights in this region as standards in evaluating the instrument constant. Furthermore, this approach eliminates the need to use "effective" [10] rather than actual molecular weight values for the carrier gases.

It should be evident from Eq. (18) and Table 1 that, in the range of molecular weights greater than the molecular weights of the carrier gases, the change in KA_1/A_2 with molecular weight becomes small as molecular weight increases. Thus small errors in K or A_1/A_2 can lead to significant errors in molecular weight calculations. Analysis of the factor KA_1/A_2 shows that at low molecular weight regions A_1/A_2 varies significantly as molecular weight changes and small errors in K are not magnified; however at high molecular weight regions A_1/A_2 tends to a constant and KA_1/A_2 becomes very sensitive to errors in K. Therefore, a K value evaluated using low molecular weight standards may, in practice, lead to large errors in estimating molecular weights of unknowns in the high molecular weight range. In contrast, a K value determined using high molecular weight standards will in general be good in estimating molecular weights of low molecular weight unknowns due to the fact that in the low molecular weight range, A_1/A_2 rather than K is the governing factor. However, as pointed out earlier, more accurate values of molecular weights of unknowns can be obtained using more than one K value applicable to the molecular weight ranges of interest.

In principle, in calibration and later in molecular weight calculations, peak areas rather than peak heights should be used. The use of peak heights can be permissible only if the peaks are not skewed and the peak base widths are identical on both channels (which imposes a strict requirement for identical columns). However, area measurements are not without ambiguities with respect to limits of integration, especially in the case of partially overlapping peaks.

POLYETHYLENE

The degradation of polyethylene by pyrolysis has an extensive literature and thus serves to demonstrate the capabilities of the present system on a comparative basis.

Literature Review

Prior to the introduction of pyrolysis-gas chromatography in 1954 [23], analyses of the products of pyrolysis were dependent upon fraction collecting and subsequent analysis of the fractions by IR and/or mass spectrometry, or by chemical methods. In one of the earlier studies, pyrolysis of a sample of polyethylene of average molecular weight 20,000, for 30 min at 475°C in vacuum, resulted in a solid residue, a waxlike fraction (which was not volatile at room temperature), and a fraction which was volatile at room temperature [24]. Only this volatile fraction was analyzed in a mass spectrometer and was reported to be composed of hydrocarbons up to heptanes. The waxlike fraction, which accounted for 95.5% of the decomposition products, was not analyzed, except for testing for its average molecular weight by a microfreezing point lowering method in camphor. The average molecular weight was reported to be 692, which corresponds to an average fragment of about 50 carbon atoms.

These initial studies were thus limited to the analysis of only the low molecular weight volatile decomposition products. A detailed review has been published [25].

The technique of gas chromatography combined with pyrolysis initiated two types of studies. Earlier applications were in the establishment of the so-called "fingerprint" pyrograms for purposes of identification of unknown polymer samples, and did not require identification of the individual peaks appearing in the pyrograms. A large number of fingerprint pyrograms of polyethylene has been published (e.g., Refs. 26-30). The pyrograms are difficult to compare due to the differences in pyrolysis and chromatographic conditions and the types of pyrolyzers utilized in various laboratories. The question of interlaboratory reproducibility in pyrolysis-gas chromatography is indeed a major concern [4, 31].

The pyrolysis-gas chromatographic techniques have been gradually directed more toward investigation of degradation mechanisms and molecular structures of polymers. These require, in particular, identification of the decomposition products.

The first detailed work which was directed to the analysis of the decomposition products of polyethylene by pyrolysis-gas chromatography was reported in 1964 [32]. A sample of polyethylene (presumably of low density) was pyrolyzed for 10 sec at 1000°C using a filament pyrolyzer. The chromatogram displayed a homologous series of triplet peaks which upon hydrogenation reduced to a homologous series of single peaks corresponding to normal hydrocarbons from C_7 to C_{17}, thus indicating that the other two peaks in the triplets corresponded to unsaturated hydrocarbons. In the same year it was reported [33] that pyrolysis of low- or high-density polyethylene samples at 690°C, using this time a reaction chamber technique, also led to a chromatogram composed of repeated triplet peaks, the resolution of which improved when capillary columns were used for separation. The peaks were indicated to correspond to C_7 to C_{27} hydrocarbons; however, verification was provided only for C_7, C_{12}, and C_{16}.

In a later study, in order to simplify identification, products of pyrolysis (for about 1 sec at 500°C) of a linear polyethylene (i.e., polymethylene) sample were hydrogenated before entry into the gas chromatographic column [34]. The peaks were identified by comparison of retention times of model compounds and found to correspond to normal hydrocarbons up to C_{13}. The pyrolysis products above C_{13} were not detected due to condensation of higher molecular weight fragments in the pyrolyzer and hydrogenation sections. A further investigation [35] addressed itself to the analysis of the nature of the triplets that were formed upon pyrolysis of polyethylene samples of different densities. Pyrolysis was carried out for 10 sec at 1000°C. By mixing the original sample with a number of normal hydrocarbons and α-olefins, the identity of two peaks of the triplets were determined to be a normal paraffin and an α-olefin. In the case of the triplet corresponding to C_{12}, the third peak was identified indirectly from chemical reaction analyses to be the α,ω-diolefin. It was then assumed that the third peak in each triplet corresponded to the α,ω-diolefin. The triplet formation was observed also in pyrolysis by electric discharge [36]; however, the peaks observed in this study were not identified. Later, in a study [37] of the pyrolysis products of polyethylene, where C_1 to C_6 hydrocarbons were identified by comparison of retention time data of model compounds, it was observed that the amount of butadiene and butene generated at a given pyrolysis temperature (in the range 450 to 900°C) depended upon the degree of branching, and an index was derived which was based on the relative peak heights of butene to determine the composition of linear and branched polyethylene samples. In the same year there appeared another report [38] on a combination of hydrogenation with pyrolysis-gas chromatography to identify decomposition products of (low- and high-density) polyethylene up to 1-hexene. In 1968 a technique was described [39] which combined thermogravimetry and gas chromatography. The philosophy of this approach differed from earlier experiments in that pyrolysis was

carried out by program heating. The chromatogram of the decomposition products of a sample of low-density polyethylene pyrolyzed at
a program heating rate of $5°C/min$ up to $500°C$ displayed a series of
evenly spaced doublet peaks which were assumed to correspond to
unsaturated and saturated straight chain hydrocarbons. This doublet
formation differs from the often reported homologous series of triplets which appears to be representative of flash pyrolysis. For
example, in the same year it was again reported [40] that flash
pyrolysis (at $680°C$ for 20 sec) of polyethylene leads to the formation
of triplets which reduce to singlets upon hydrogenation. The range of
detectable fragments was claimed to be up to C_{36} even though, in the
published pyrogram, the assignment is only up to C_{24} (from C_5). By
modification of the chromatographic conditions, it was reported in the
same paper that hydrocarbons up to C_{50} could be detected. However,
the pyrogram shows assignments only from C_{10} to C_{40}. Two additional
studies appeared in the same year [41, 42]. In a detailed study of the
volatile decomposition products up to C_9, 30 peaks were detected for
hydrocarbons from C_1 to C_6 and were identified by comparing
retention times [41]. Samples were pyrolyzed in vacuum isothermally
at various temperatures in the range of 375 to $425°C$. The same
authors also investigated the thermal decomposition products of polymethylene and reported that in the chromatogram in which the products
from C_1 to C_{16} hydrocarbons were analyzed, the low molecular weight
fragments below C_8 were doublets consisting of α-olefin and n-alkane;
however, above $n\text{-}C_8$ hydrocarbons, a small peak appeared before the
α-olefin peak, resulting in a series of triplet peaks In a later study
reported in 1970 [43], hydrocarbons up to C_{12} were analyzed. More
recent studies [44, 45] have concerned themselves with the structure
of ethylene-propylene copolymers with respect to the determination of
sequence length distribution in these copolymers by pyrolysis-gas
chromatography.

The formation of the saturated and unsaturated hydrocarbons reported in these studies has been interpreted in terms of a random chain
scission process involving a free radical mechanism [25, 34, 35, 41, 42,
46]. Considerable intramolecular radical transfer, especially,to the 5th
carbon by a coiling mechanism, is operative [34, 41, 42] under mild
conditions.

Present Results

A 20-mg sample of low-density polyethylene ($\rho = 0.925$, $\overline{M}_w = 140,000$,
$\overline{M}_n = 16,000$, 13 methyl groups/1000 C-atoms) was pyrolyzed in the
present system by program heating (at $20°C/min$) from room temperature to $600°C$, and the fraction volatilizing between 300 and $600°C$ was
analyzed in the mass chromatograph.

The thermal conductivity cell response during pyrolysis and also the complementary results of thermogravimetric and differential thermal analyses (obtained at same heating rate but in a nitrogen atmosphere, using a Du Pont 950 TGA and a 900 DTA unit) are shown in Fig. 3. This documents the important thermal history before and during pyrolysis.

The mass chromatograph output which is shown in Fig. 4 displays a series of regularly spaced (doublet) peaks. Calculations of the molecular weights of each prominant peak indicate that the molecular weights of the compounds corresponding to these peaks are comparable to those of C_2 to C_{26} hydrocarbons (Table 2) and have been so-designated in the figure. The constituents more volatile than C_3 are not discussed herein. The molecular weight calculations further indicate that the first peak in each doublet has a molecular weight comparable to that of an unsaturated hydrocarbon (with one double bond) and the second peak to that of a saturated hydrocarbon. The doublets merge beyond C_{16} and below C_5; they are not resolved under the present chromatographic conditions. Each peak is probably a mixture of a saturated hydrocarbon of the indicated carbon number.

Discussion

A direct comparison of the present results with those reported in the literature is not easy due to differences both in the pyrolysis and the chromatographic conditions. However, certain similarities exist.

As mentioned earlier, there has been only one published study in which pyrolysis of polyethylene was carried out by program heating rather than by flash heating and the decomposition products were analyzed by gas chromatography [39]. In that study, also, the pyrogram displayed a homologous series of doublet peaks similar to the present observations. In contrast, flash-pyrolysis experiments have been reported to lead to triplet formation [32, 33, 35, 36, 40, 44]. A close examination of Fig. 4 indicates that starting with the doublet corresponding to the C_9 hydrocarbons, a small peak starts to appear as a shoulder before the alkene peak, which is similar to a phenomenon reported in an isothermal-pyrolysis experiment with polymethylene [42].

The relative amounts of alkadiene produced are dependent upon the mechanism of degration. Homolytic chain scission followed by simultaneous saturation and unsaturation at the site of scission predicts a yield of n-alkane, 1-alkene, and alkadiene in the mole ratio 1:2:1

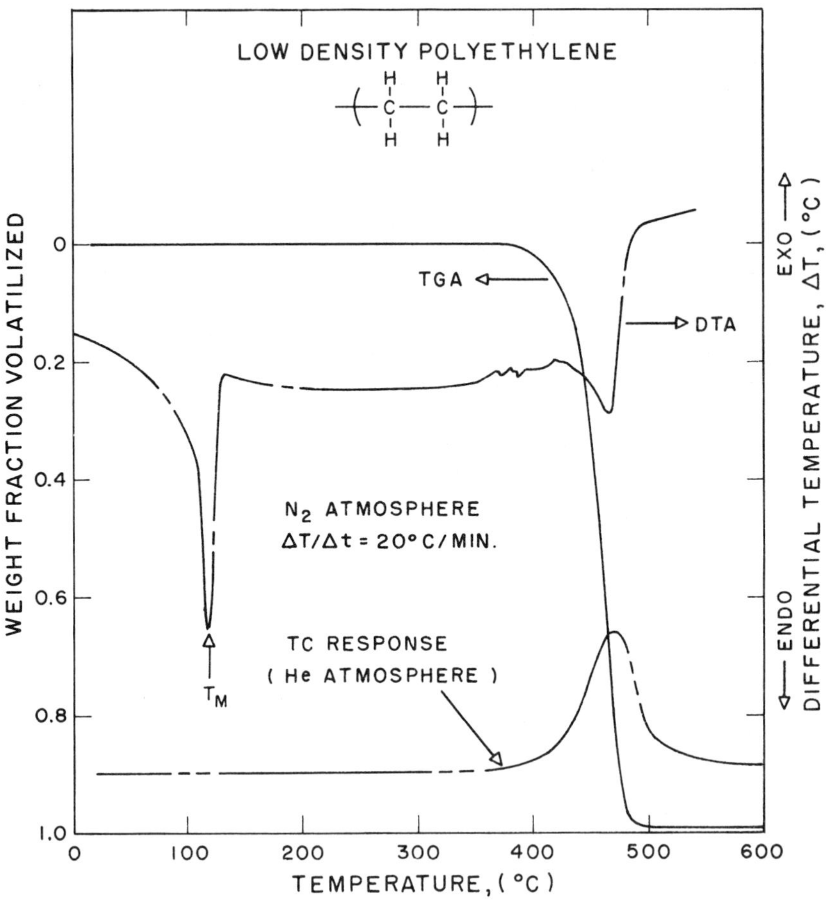

FIG. 3. Thermal history before and during pyrolysis of low-density polyethylene ($\rho = 0.925$, $\overline{M}_w = 140{,}000$, $\overline{M}_n = 16{,}000$, 13 methyl groups/ 1000 C-atoms).

[35, 42], indicating that the quantity of alkadiene produced should be as much as n-alkane. A predominantly intermolecular radical transfer process, however, predicts [42] that the amount of alkadiene produced should be greater than that of n-alkane. In contrast, a predominantly intramolecular radical transfer process by back-biting predicts that the amount of alkadiene should be very small, being zero if it is the only process that is operative [42].

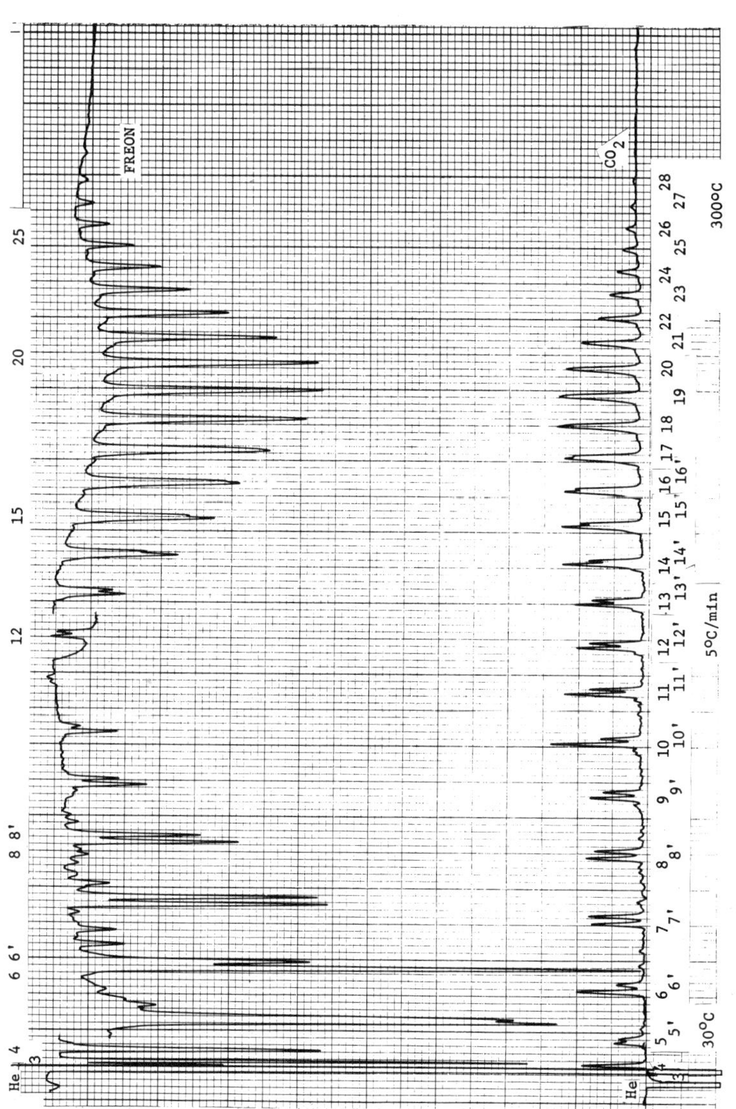

FIG. 4. Mass chromatogram of the pyrolysis products of low-density polyethylene ($\rho = 0.925$, $\overline{M}_W = 140,000$, $\overline{M}_n = 16,000$, 13 methyl groups/1000 C-atoms). A 20-mg sample was pyrolyzed by program-heating (at 20°C/min) in helium atmosphere, and the fraction from 300 to 600°C was analyzed. The columns were operated isothermally at 30°C until C_4 peaks eluted and then program-heated (at 5°C/min) to 300°C and held isothermally at 300°C. The peak attenuations were ×8 except for the C_3 and C_4 peaks in the Freon-115 channel for which the settings were ×64. After elution of C_{12} peaks, the polarity of the Freon-115 detector response was reversed to result in downward peaks for constituents of higher molecular weight. For a comparison of the retention times, see Fig. 2.

TABLE 2. Molecular Weights of Degradation Products of Polyethylene

Peak number (x)	Response ratio A_1/A_2 (Freon/CO_2)	Molecular weight (calcd)	Molecular weight of saturated x-hydrocarbon[a]	Tentative assignments C number (n, saturated; $n^=$, olefin)
3	406	42.67	44.09	$3^=$,3
4	-39.13	56.20	58.12	$4^=$,4
5	-14.32	71.97		5
5'	-15.19	70.74	72.15	$5^=$
6	- 8.20	85.08		$6^=$
6'	- 8.08	85.46	86.18	6
7	- 4.87	99.16		$7^=$
7'	- 4.41	101.91	100.21	7
8	- 3.08	111.58		$8^=$
8'	- 2.78	114.21	114.23	8
9	- 1.61	126.95		$9^=$
9'	- 1.44	129.19	128.26	9
10	- 0.64	141.60		$10^=$
10'	- 0.57	142.86	142.29	10
11	0.074	156.03		$11^=$
11'	0.174	158.22	156.31	11

12	0.60	168.59		$12^{=}$
12'	0.66	170.21	170.34	12
13	1.00	180.20		$13^{=}$
13'	1.11	183.78	184.27	13
14	1.43	195.36		$14^{=}$
14'	1.51	198.57	198.40	14
15	1.74	208.58		$15^{=}$
15'	1.86	214.34	212.31	15
16	2.02	222.67		$16^{=}$
16'	2.14	229.47	226.45	16
17	2.33	241.37	240.35	$17^{=}$, 17
18	2.49	252.64	254.50	$18^{=}$, 18
19	2.72	267.12	268.39	$19^{=}$, 19
20	2.89	287.40	282.56	$20^{=}$, 20
21	2.96	294.71	296.43	$21^{=}$, 21
22	3.08	308.32	310.59	$22^{=}$, 22
23	3.17	319.53	324.47	$23^{=}$, 23
24	3.29	336.06	338.66	$24^{=}$, 24
25	3.38	349.81	352.51	$25^{=}$, 25
26	3.50	370.30	366.72	$26^{=}$, 26

[a]CRC Handbook of Chemistry and Physics, 48th ed.

Evidence for predominant intramolecular radical transfer by back-biting in the degradation of polyethylene can thus be found [42] in the experimental observations of relative amounts of alkadiene formation In Fig. 4 the 1-alkene peak is larger than the n-alkane peak in each doublet but only small amounts of alkadienes are detectable (beyond C_8). This suggests that in the program-heating mode the radicals that are formed apparently have enough time to coil and abstract a hydrogen intramolecularly. In flash-pyrolysis experiments, however, radicals are subject to a large number of secondary reactions and do not have time for intramolecular transfer reactions.

It is interesting to note that the shape of the chromatogram (Fig. 4) in the C_{12} to C_{28} region is pyramidal and peaks at about C_{19}. A similar phenomenon was reported in pyrograms of both high- and low-density polyethylene [40], even though the experiments were carried out by flash pyrolysis. It was observed that low-density polyethylene showed a maximum in the vicinity of C_{18} to C_{19}, in agreement with the present observations, whereas high-density polyethylene showed two maxima occurring around C_{22} to C_{23} and C_{30} to C_{32}. (This difference was indicated to be useful in differentiation between different polyethylene samples.) The small peaks that are observed between the doublets (in Fig. 4) presumably result from branching in low-density polyethylene and thus may correspond to branched hydrocarbons. The formation of such branched hydrocarbons is reported to take place to a lesser extent in high-density polyethylene, and this observation also has been suggested for differentiation between polyethylene samples [40].

It is evident from this example with polyethylene and studies conducted with polyisobutylene, polypropylene, polystyrene, and a series of poly-sulfones [47] that the present technique of pyrolyzer/thermal conductivity probe/mass chromatograph gives the molecular weights of constituents of pyrolysis with useful accuracy and probably with greater ease than conventional methods. The accuracy of the results can no doubt be improved with improved resolution of the peaks. The present system offers certain other advantages. The program-heating capability of the pyrolyzer is most suitable for studying mechanisms of the onset of thermal decomposition and for following its progress. Furthermore, pyrolysis is conducted with minimum thermal energy which is essential to minimize secondary reactions. Use of program-heating pyrolysis has been previously reported only in a few studies where a thermogravimetric analyzer was combined with a gas chromatograph [39, 48] or a mass spectrometer [49]. The importance of this approach is more evident if it is realized that in most cases in flash pyrolysis, pyrolysis is completed well below the temperature of the heat source [50]. The uncertainties with respect to the temperature of pyrolysis can thus be resolved by utilizing program heating, and interlaboratory reproducibility can be achieved. Perhaps one of the reasons for not using program heating in earlier studies was because complete pyrolysis is not achieved rapidly, which would mean introduction of pyrolysis products onto the column over an extended period of time, and the question of broadening of peaks in the chromatogram would arise. The utilization of traps in the present system eliminates this problem. The decomposition products can be collected for any period of time and later can be released and introduced directly to the gas chromatographic column as a sharp plug by flash heating.

The thermal conductivity cell detector provides the flexibility for selective trapping and is most useful in monitoring the progress of decomposition under experimental pyrolysis conditions. However, the detector oven cannot be maintained above 250°C, and thus introduces a limitation for analysis of high molecular weight decomposition fragments. Only those constituents that are volatile at the detector oven temperature are carried to the pyrolyzer trap (Fig. 1), and later released and analyzed with the mass chromatograph.

The utilization of gas density balance detectors provides another unique feature to the present system in that oxidative studies can be conducted since constituents analyzed in the mass chromatograph never come in contact with the filaments of the detectors (Fig. 1).

The effect of temperature programming in pyrolysis and the nature of the decomposition products generated in different temperature ranges in the course of decomposition of polyethylene and other polyolefins are being investigated [51]. Finally, it is interesting to note that the present pyrogram shown in Fig. 4 covers the widest range of compounds (from C_3 to C_{26}) with determined molecular weights that has been published in a single pyrogram of polyethylene.

ACKNOWLEDGMENTS

This work was supported by the Chemistry Branch of the Office of Naval Research (Contract N00014-67-A-0151-0024) and the Textile Research Institute, Princeton, New Jersey.

REFERENCES

[1] R. W. McKinney, "Pryolysis-Gas Chromatography," in Ancillary Techniques of Gas Chromatography" (E. McFadden, ed.), Wiley, New York, 1969, pp. 55-87.

[2] M. P. Stevens, Characterization and Analysis of Polymers by Gas Chromatography, Dekker, New York, 1969.

[3] G. M. Brauer, "Pyrolysis-Gas Chromatographic Techniques for Polymer Identification" in Thermal Characterization Techniques Techniques and Methods of Polymer Evaluation Series, Vol. 2), (P. E. Slade, Jr. and L. T. Jenkins, eds.), Dekker, New York, 1970, pp. 41-105.

[4] J. Q. Walker, Chromatographia, 5, 547 (1972).

[5] S. G. Perry, Chromatogr. Rev., 9, 1 (1967).

[6] A. B. Littlewood, Chromatographia, 1, 37, 133, 223 (1968).

[7] D. A. Leathard and B. C. Shurlock, Identification Techniques in Gas Chromatography, Wiley (Interscience), New York, 1970.

[8] D. G. Paul and G. E. Umbreit, Res./Develop., 21, 18 (1970).

[9] C. E. Bennett et al., Amer. Lab., 3, 67 (1971).

[10] R. S. Swingle, Ind Res., 14(2), 40 (1972).

[11] Chromalytics Corporation, Route 82, Unionville, Pennsylvania.

[12] A. J. P. Martin and A. T. James, Biochem. J., 63, 138 (1956).

[13] A. Liberti, L. Conti, and V. Crescenzi, Nature, 178, 1067 (1956).

[14] C. S. G. Phillips and P. L. Timms, J. Chromatogr. 5, 131 (1961).

[15] A. G. Nerheim, Anal. Chem., 35(11), 1640 (1963).

[16] Gow-Mac Instrument Co., 100 Kings Road, Madison, New Jersey.

[17] R. Quillet, J. Chromatogr. Sci., 8, 405 (1970).

[18] J. P. Parsons, Anal. Chem., 36(9), 1849 (1964).

[19] C. L. Guillemin and F. Auricourt, J. Gas Chromatogr., 1, 24 (1963).

[20] C. L. Guillemin and F. Auricourt, Ibid., 2, 156 (1964).

[21] C. L. Guillemin, F. Auricourt, and P. Blaise, Ibid., 4, 338 (1966).

[22] J. T. Walsh and D. M. Rosie, Ibid., 5, 232 (1967).

[23] W. H. T. Davison, S. Slaney, and A. L. Wragg, Chem. Ind. (London), 1954, 1356.

[24] S. L. Madorsky et al., J. Polym. Sci., 4, 639 (1949).

[25] S. L. Madorsky, Thermal Degradation of Organic Polymers, (Polymer Reviews, Vol. 7) Wiley, New York, 1964, Chap. 4.

[26] W. H. Pariss and P. D. Holland, Brit. Plastics, 33, 372 (1960).

[27] J. G. Cobler and E. P. Samsel, Soc. Plastics Eng. SPE, Trans., 2, 145 (1962).

[28] B. C. Cox and B. Ellis, Anal. Chem., 36(1), 90 (1964).

[29] B. Groten, Ibid., 36(7), 1206 (1964).

[30] F. W. Willmott, J. Chromatogr. Sci., 7, 101 (1969).

[31] N. B. Coupe, C. E. R. Jones, and S. G. Perry, J. Chromatogr. 47, 291 (1970).

[32] B. Kolb and K. H. Kaiser, J. Gas Chromatogr., 2, 233 (1964).

[33] E. W. Cieplinski, et. al., Z. Anal. Chem., 205, 357 (1964).

[34] J. van Schooten and J. K. Evenhuis, Polymer, 6, 343 (1965).

[35] B. Kolb et al., Z. Anal. Chem., 209, 302 (1965).

[36] J. C. Sternberg and R. L. Litle, Anal. Chem., 38(2), 321 (1966).

[37] D. Deur-Siftar, J. Gas Chromatogr., 5, 72 (1967).

[38] S. K. Yasuda, J. Chromatogr., 27, 72 (1967).

[39] J. Chiu, Anal. Chem., 40(10), 1516 (1968).

[40] L. Michajlov, P. Zugenmaier, and H.-J. Cantow, Polymer, 9, 325 (1968).

[41] Y. Tsuchia and K. Sumi, J. Polym. Sci., A-1, 6, 415 (1968).

[42] Y. Tsuchia and K. Sumi, Ibid., B, 6, 357 (1968).

[43] A. A. Buniyat-Zade et. al., Polym. Sci. USSR, A12, 2825 (1970).

[44] L. Michajlov, H.-J. Cantow, and P. Zugenmaier, Polymer, 12, 70 (1971).

[45] H. Seno, S. Tsuge, and T. Takeuchi, Makromol. Chem., 161, 195 (1972).
[46] L. A. Wall et. al., J. Amer. Chem. Soc., 76, 3430 (1954).
[47] E. Kiran and J. K. Gillham, Soc. Plastics Eng. SPE, Tech. Papers, 19, 502 (1973).
[48] G. Blandenet, Chromatographia, 2, 184 (1965).
[49] F. Zitomer, Anal. Chem., 40(7), 1091 (1968).
[50] F. Farré-Ruis and G. Guiochon, Ibid., 40(6), 998 (1968).
[51] E. Kiran, Ph.D. Thesis, Princeton University, 1973.

Note added in proof. A detailed analysis of the hydrodynamics of operation of the gas density balance is now available [51]. A more extensive discussion of the technique of pyrolysis-molecular weight chromatography and its applications to the study of the thermal degradation of polyolefins, polyalkylmethacrylates, and olefin/sulfone copolymers are also presented in Ref. 51.

Numbers in brackets are reference numbers and indicate that an author
work is referred to although his name is not cited in the text. Underlin(
numbers give the page on which the complete reference is listed.